Verilog Coding for Logic Synthesis

Verilog Coding for Logic Synthesis

WENG FOOK LEE

A JOHN WILEY & SONS, INC., PUBLICATION

Published by John Wiley & Sons, Inc., Hoboken, New Jersey.
Published simultaneously in Canada.

For general information on our other products and services please contact our Customer Care Department within the U.S. at 877-762-2974, outside the U.S. at 317-572-3993 or fax 317-572-4002.

Wiley also publishes its books in a variety of electronic formats. Some content that appears in print, however, may not be available in electronic format.

Library of Congress Cataloging-in-Publication Data:

Library of Congress Cataloging-in-Publication Data is available.
ISBN 0-471-42976-7

10 9 8 7 6 5 4 3 2 1

Dedicated to my mother
for all her sacrifices.

Contents

Table of Figures

Table Of Examples

List Of Tables

Preface

The complexity of integrated circuit (IC) chips has increased tremendously over the past 10 years. In the 1980s, designing an IC chip with several million transistors was simply unimaginable. Today, it is common to have several million transistors on an IC chip. This increase in IC chip complexity is mainly the result of integration of many functions into a single IC chip. With this fundamental change, the conventional method of schematic capture used in IC design became an obstacle to design engineers. It became extremely difficult for design engineers to "hand draw" the large amounts of schematics necessary to achieve the required functionality. Furthermore, IC chips are pushed onto the market at a very fast pace, creating a small time-to-market window. Designers are under constant pressure to design more complex IC chips at a faster rate.

Imagine design engineers needing to hand draw millions of transistors in their schematic! The task was simply impossible. A more efficient and productive method was needed to allow designers to create schematics with large numbers of logic gates within a reasonable timeframe. This lead to the development of ***h***ardware ***d***escription ***l***anguage (HDL).

This new method allows a designer to code the logic functionality of a circuit in HDL. The code is then synthesized into logic gates using a synthesis tool. A common synthesis tool used in the industry is Synopsys's Design Compiler. (To learn how to use Synopsys's Design Compiler and to write very high-speed integrated hardware description language [VHDL] code, refer to ***VHDL Coding and Logic Synthesis with Synopsys***, by ***Weng Fook Lee, Academic Press***.)

There are two types of HDL used in the industry: Verilog and VHDL. This book only addresses Verilog.

This book is written specifically for students and engineers learning to write synthesizable Verilog code. Chapter 1 introduces the use of VHDL and Verilog. Chapter 2 describes application-specific IC (ASIC) design flow. Flow charts and descriptions are given to help the reader better understand ASIC design flow.

Chapter 3 discusses basic concepts of Verilog coding. This chapter shows the reader how numbers, comments, and Verilog data types and strengths can be used in Verilog coding. Use of Verilog gate-level primitives and user-defined primitives are also discussed.

Chapter 4 describes the common practices and coding style used when coding for synthesis. Naming convention, design partitioning, effects of timing loops, clock generation, reset usage, and sensitivity list are covered in this chapter. Verilog concepts of blocking and nonblocking statements are discussed in detail. Examples and waveforms are provided throughout to help the reader understand these concepts. Chapter 4 also gives examples of common coding style for Verilog operators. The concepts of latch inference, coding of memory array, and the state machine are also included. The state machine design example consists of design specification, state diagrams to show the functionality of the state machine, synthesizable Verilog code for the state machine, and test benches to verify the functionality of the state machine.

Chapter 5 shows the reader how a design project for a programmable timer is implemented. This chapter starts with a specification for a programmable timer. It then proceeds to show the reader how a microarchitecture can be derived from the specification. Flow charts are shown to help the reader understand the functionality that is required. Verilog code and Verilog test benches are included to show how the design example can be simulated and verified. Waveforms of output results are discussed.

Chapter 6 shows the design of a programmable logic block for peripheral interface (similar to the industry's 8255 PPI [programmable peripheral interface]). This chapter begins with the specification of the design. Microarchitecture of the design is discussed and flow charts are shown to help the reader understand the required functionality of the design. Synthesizable Verilog code for the design is shown with test benches for verification of different functionality of the design. Waveforms for output results are discussed.

This book gives many examples and is written with practicality in mind. It has 91 examples to help the reader understand the concepts and coding style that are being discussed. It begins with simple Verilog coding and progresses to complex real-life design examples. Chapter 4 shows a state machine design example of an intelligent traffic light system. Chapter 5 shows a design example of a programmable timer, beginning with product specification, microarchitecture definition, Verilog coding, and verification. This design example also shows the reader how Verilog code can be written and verified but cannot be synthesized into the required circuit.

To help the reader gain a better understanding of how these real-life design examples are achieved, flow charts, waveforms, and detailed explanations of simulation results are included.

Acknowledgments

This book would not have been possible without the help of many people. I would like to thank Mike Van Buskirk, Colin Bill, Ken Cheong Cheah, Ed Bautista, Santosh Yachareni, Tim Chen, Yu Cheng, Murni, Boon Weng, Keith Wong, Azrul Halim, Azydee Hamid, Chun Keat Lee, Sun Chong See, Soo Me, Mona Chee, Dianne and Larry, and, of course, the staff at John Wiley & Sons.

Trademarks

HDL Designer and Modelsim are trademarks of Mentor Graphics, Inc.
Visual HDL is a trademark of Summit Design, Inc.
Design Compiler, VCS, and Scirocco are trademarks of Synopsys, Inc.
Verilog XL, NC Sim, and Ambit are trademarks of Cadence, Inc.

CHAPTER ONE

Introduction

Since the early 1980s, when schematic capture was introduced as an efficient way to design very large-scale integration (VLSI) circuits, it has been the design method of choice for designers in the world of VLSI design.

However, the use of this method reached its limits in the early 1990s, as more and more logic functionality and features were integrated onto a single chip. Today, most application-specific integrated circuit (ASIC) chips consist of no fewer than one million transistors. Designing circuits this large using the method of schematic capture is time consuming and is no longer efficient. Therefore, a more efficient manner of design was required. This new method had to increase the designers' efficiency and allow ease of design, even when dealing with large circuits.

From this requirement arose the wide acceptance of HDL (hardware description language). HDL allows a designer to describe the functionality of a required logic circuit in a language that is easy to understand. The description is then simulated using test benches. After the HDL description is verified for logic functionality, it is synthesized to logic gates by using synthesis tools.

This method helps a designer to design a circuit in a shorter timeframe. The savings in design time is achieved because the designer need not be concerned with the intricate complexities that exist in a particular circuit, but instead is focused on the functionality that is required. This new method of design has been widely adopted today in the field of ASIC design. It allows designers to design large numbers of logic gates to implement logic features and functionality that are required on an ASIC chip.

Verilog Coding for Logic Synthesis, edited by Weng Fook Lee
ISBN 0-471-429767

The most widely used HDLs in the ASIC industry are Verilog and VHDL (very high-speed integrated circuit hardware description language). Each have advantages and disadvantages. The coding styles for these languages have some similarities as well as differences.

CHAPTER TWO

ASIC Design Flow

Application-specific integrated circuit (ASIC) design is based on a design flow that uses hardware description language (HDL). Most electronic design automation (EDA) tools used for ASIC flow are compatible with both Verilog and very high speed integrated circuit hardware description language (VHDL).

In this flow, the design and implementation of a logic circuit are coded in either Verilog or VHDL. Simulation is performed to check its functionality. This is followed by synthesis. Synthesis is a process of converting HDL to logic gates. After synthesis, the next step is APR (auto-place-route). APR is explained in more detail in Section 2.6.

Figure 2.1 shows a diagram of an ASIC design flow, beginning with specification of an ASIC design to register transfer level (RTL) coding and, finally, to tapeout.

2.1 SPECIFICATION

Figure 2.2 indicates the beginning of the ASIC flow: the specification of a design. This is Step 1 of an ASIC design flow. The design of an ASIC chip begins here.

Specification is the most important portion of an ASIC design flow. In this step, the features and functionalities of an ASIC chip are defined. Chip planning is also performed in this step.

During this process, architecture and microarchitecture are derived from the required features and functionalities. This derivation is especially impor-

Verilog Coding for Logic Synthesis, edited by Weng Fook Lee
ISBN 0-471-429767

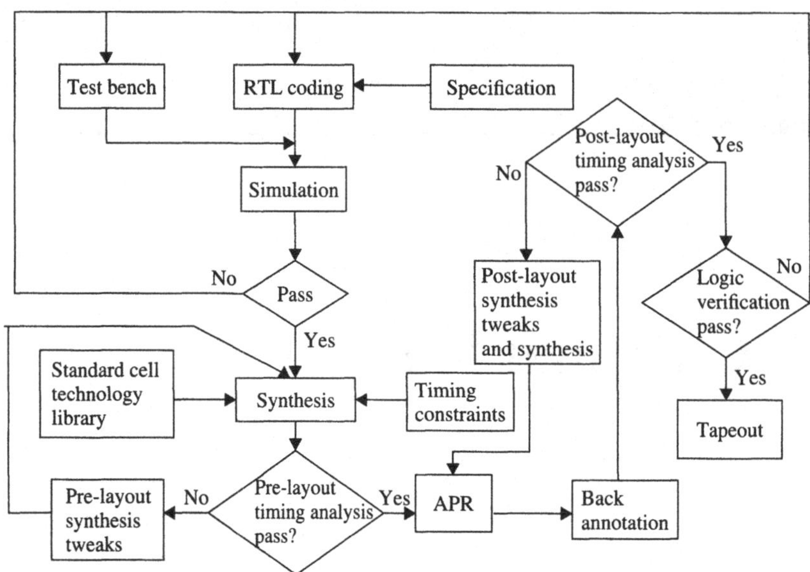

FIGURE 2.1. Diagram showing an ASIC design flow. Sections 2.1 to 2.9 explain each section of the ASIC flow in detail.

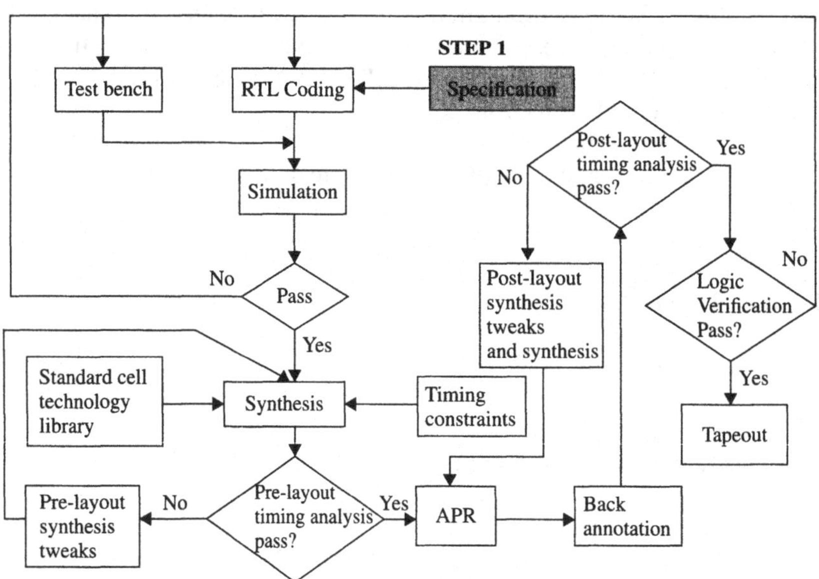

FIGURE 2.2. Diagram indicating Step 1 of an ASIC design flow: specification.

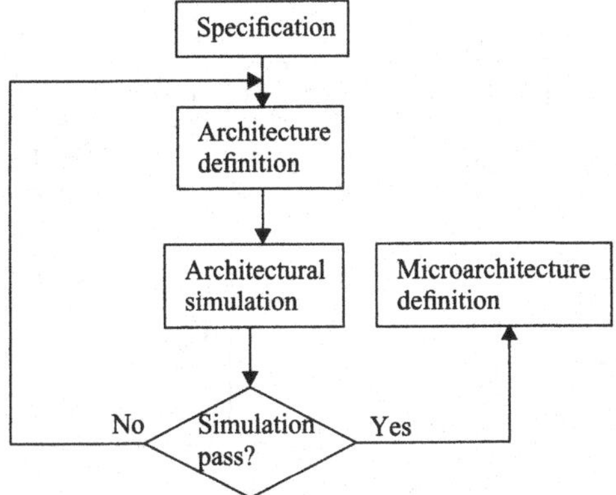

FIGURE 2.3. Diagram showing the definition of architecture and microarchitecture.

tant, as the architecture of a design plays an important role in determining the performance capabilities and silicon area utilization.

Figure 2.3 shows the process involved in defining the architecture and microarchitecture of a design. Specification contains a list of all features and functionalities required in the design. These include power consumption, voltage references, timing restrictions, and performance criteria. From this list, the chip architecture can be drafted. This defined architecture must take into consideration all required timing, voltage, and speed/performance of the design. Architectural simulations need to be performed on the drafted architecture to ensure that it meets the required specification.

During architecture simulations, the architectural definition will have to be changed if the simulation result shows it cannot meet any requirements in the specification. When all the requirements are met, this architecture is said to meet the required specifications. From here, a microarchitecture is drafted and defined to allow execution of the architecture from a design standpoint.

The microarchitecture is the key point that enables the design phase. A microarchitecture interfaces the design's architecture and circuit. It also allows transformation of an architectural concept into possible design implementation.

2.2 RTL CODING

Figure 2.4 shows Step 2 of the ASIC design flow. This is the beginning of the design phase. The microarchitecture is transformed into a design by converting it into RTL code.

As shown in Section 2.1 (Step 1 of the ASIC design flow), architecture and microarchitecture are derived from specification. In Step 2, the microarchitecture, which is the implementation of the design, is coded in synthesizable RTL.

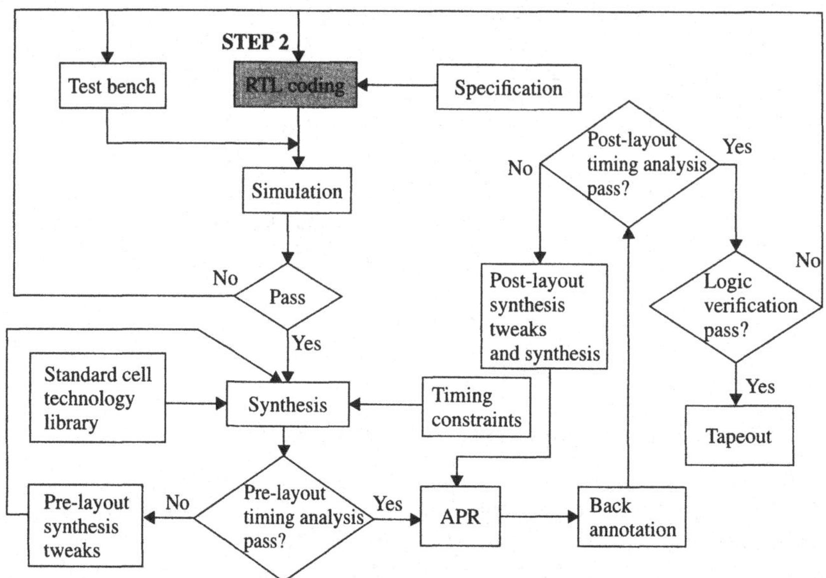

FIGURE 2.4. Diagram indicating Step 2 of an ASIC design flow: RTL coding.

There are several ways to obtain the RTL code. Some designers use graphical entry tools like Summit Design's Visual HDL or Mentor Graphics HDL Designer. These graphical entry tools allow designers to use bubble diagrams, flow charts, or truth table to implement the microarchitecture, which subsequently generate the RTL code either in Verilog or VHDL. However, some designers prefer writing the RTL code rather than using a graphical entry tool. Both approaches end in the same result: synthesizable RTL code that describes logic functionality of the specification.

2.2.1 Types of Verilog Code: RTL, Behavioral, and Structural

Section 2.2 discusses RTL coding. In Verilog language, there are three types of Verilog code. For most cases of synthesis, synthesizable RTL code is used. Table 2.1 lists the differences and usage of each of the types of Verilog code.

2.3 TEST BENCH AND SIMULATION

Figure 2.5 shows Step 3 in the ASIC design flow, which involves creation of test benches. These are used to simulate the RTL code.

A test bench is basically a wraparound environment surrounding a design, which enables the design to be simulated. It injects a specified set of stimulus

TABLE 2.1. The three types of Verilog code

<table>
<tr><th>RTL</th><th>Behavioral</th><th>Structural</th></tr>
<tr><td>RTL coding, or register transfer level, is most commonly used to describe the functionality of a design for synthesis. It is also descriptive in nature, similar to behavioral Verilog. However, it only uses a subset of Verilog syntax, as not all Verilog syntax is synthesizable. RTL coding can be viewed as more descriptive than structural Verilog but less descriptive compared with behavioral Verilog.</td><td>Behavioral coding is used to describe a "black box" design whereby the output of the design is specified for a certain input pattern. Behavioral code mimics the functionality and behavior of the "black box" design. It is normally used for system-level testing.</td><td>Structural Verilog coding has a data type structure that defines the different components and their interconnects present in a design. It represents a netlist of a design. Structural Verilog is normally used when passing netlist information of a design between design tools. For example, upon completion of synthesis, the netlist of a design is passed to APR (refer to Section 2.6 for explanation of APR) using structural Verilog.</td></tr>
<tr><td>module RTL (inputA, inputB, inputC, inputD, outputA);

input inputA, inputB, inputC, inputD;
output outputA;
reg outputA;
always @ (inputA or inputB or inputC or inputD)
begin
if (inputA & inputB & ~inputD)
outputA = inputC;
else if (inputA & inputD & ~inputC)
outputA = inputB;
else
outputA = 0;
end
endmodule</td><td>module behavior (inputA, inputB, inputC, inputD, outputA);

input inputA, inputB, inputC, inputD;
output outputA;
reg outputA;
always @ (inputA or inputB or inputC or inputD)
begin
if (inputA & inputB & ~inputD)
outputA = #5 inputC;
else if (inputA & inputD & ~inputC)
outputA = #3 inputB;
else if ((inputA == 1'bx) | (inputB == 1'bx) | (inputC == 1'bx) | (inputD == 1'bz))
outputA = #7 1'bx;
else if ((inputA == 1'bz) | (inputB == 1'bZ))
outputA = #7 1'bZ;
else
outputA = #3 0;
end
endmodule</td><td>module structural (inputA, inputB, inputC, inputD, outputA);

input inputA, inputB, inputC, inputD;
output outputA;
wire n30;
AN3 U8 (.A(inputA), .B(n30), .C(inputB), .Z(outputA));
EO U9 (.A(inputD), .B(inputC), .Z(n30));
endmodule</td></tr>
</table>

TABLE 2.1. (Continued)

RTL	Behavioral	Structural
Referring to the Verilog code shown, when simulated or synthesized, both the RTL and structural Verilog will yield the same functionality. Behavioral Verilog, however, is not synthesizable.		

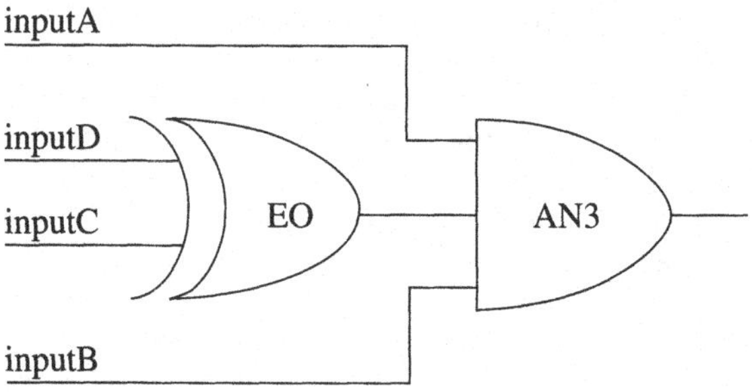

Synthesized logic for RTL Verilog and structural Verilog

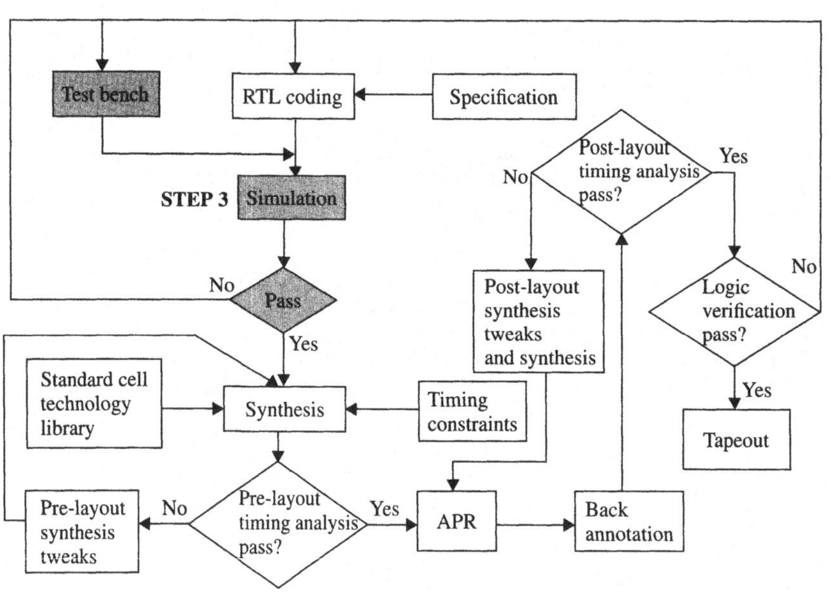

FIGURE 2.5. Diagram indicating Step 3 of an ASIC design flow: test bench and simulation.

into the inputs of the design, check/view the output of the design to ensure the design's output patterns/waveforms match designer's expectations.

RTL code and the test bench are simulated using HDL simulators. If the RTL code is written in Verilog, a Verilog simulator is required. If the RTL code is written in VHDL, a VHDL simulator is required. Cadence's Verilog XL,

Synopsys's VCS, and Mentor Graphic's Modelsim are among some of the Verilog simulators used. Cadence's NCSim and Mentor Graphic's Modelsim are capable of simulating both Verilog and VHDL. Synopsys's Scirocco is an example of a VHDL simulator. Apart from these simulators, there are many other VHDL and Verilog simulators. Whichever simulator is used, the end result is the verification of the RTL code of the design based on the test bench that is written.

If the designer finds the output patterns/waveforms during simulation do not match what he or she expects, the design needs to be debugged. A non-matching design output can be caused by a faulty test bench or a bug in the RTL code. The designer needs to identify and fix the error by fixing the test bench (if the test bench is faulty) or making changes to the RTL code (if the error is caused by a bug in the RTL code).

Upon completion of the change, the designer will rerun the simulation. This is iterated in a loop until the designer is satisfied with the simulation results. This means that the RTL code correctly describes the required logical behavior of the design.

2.4 SYNTHESIS

Figure 2.6 shows Step 4 of the ASIC design flow, which is synthesis. In this step, the RTL code is synthesized. This is a process whereby the RTL code is converted into logic gates. The logic gates synthesized will have the same logic functionality as described in the RTL code.

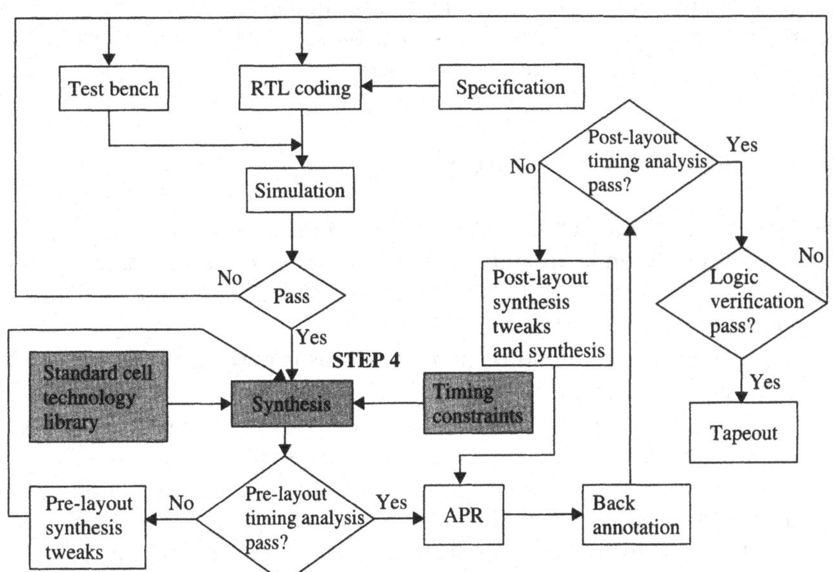

FIGURE 2.6. Diagram indicating Step 4 of an ASIC design flow: synthesis.

In Step 4, a synthesis tool is required to convert the RTL code to logic gates. More common tools used in the ASIC industry include Synopsys's Design Compiler and Cadence's Ambit.

The synthesis process requires two other input files to make the conversion from RTL to logic gates. The first input file that the synthesis tool must have before making the conversion is the "technology library" file. It is a library file that contains standard cells. During the synthesis process, the logic functionality of the RTL code is converted to logic gates using the available standard cells in the technology library. The second input file, "constraints file," helps to determine the optimization of the logic being synthesized. This file normally consists of information like timing and loading requirements and optimization algorithms that the synthesis tool needs to optimize the logic, and even possibly design rule requirements that need to be considered during synthesis.

Step 4 is a very important step in the ASIC design flow. This step ensures that synthesis tweaks are performed to obtain the most optimal results possible, should the design not meet the specified performance or area.

If, upon final optimization, the required performance or area utilization is still not within acceptable boundaries, the designer must reconsider the microarchitecture as well as architectural definitions of the design. The designer must re-evaluate to ensure the specified architecture and microarchitecture can meet the required performance and area. If the requirements cannot be met with the current architecture or microarchitecture, the designer will have to consider changing the definition of the architecture or microarchitecture. This is undesirable, as changing the architecture or microarchitecture can potentially bring the design phase back to the early stages of Step 1 of the ASIC design flow (specification). If by changing the architecture and microarchitecture definition the design is still unable to provide the kind of performance or area utilization required, the designer must resort to the possibility of changing the specification itself.

2.5 PRE-LAYOUT TIMING ANALYSIS

When synthesis is completed in Step 4, the synthesized database along with the timing information from Step 4 is used to perform a static timing analysis (Step 5). In Step 5, timing analysis is pre-layout, because the database is without any layout information (Fig. 2.7).

A timing model is built and its timing analysis is performed on the design. Normally, the timing analysis is performed across all corners with different voltages and temperatures. This is to catch any possible timing violations in the design when used across specified temperature and voltage range.

Any timing violation caught, for example, setup and hold time violations, will have to be fixed by the designer. The most common way of fixing these timing violations is to create synthesis tweaks to fix those paths that are failing timing.

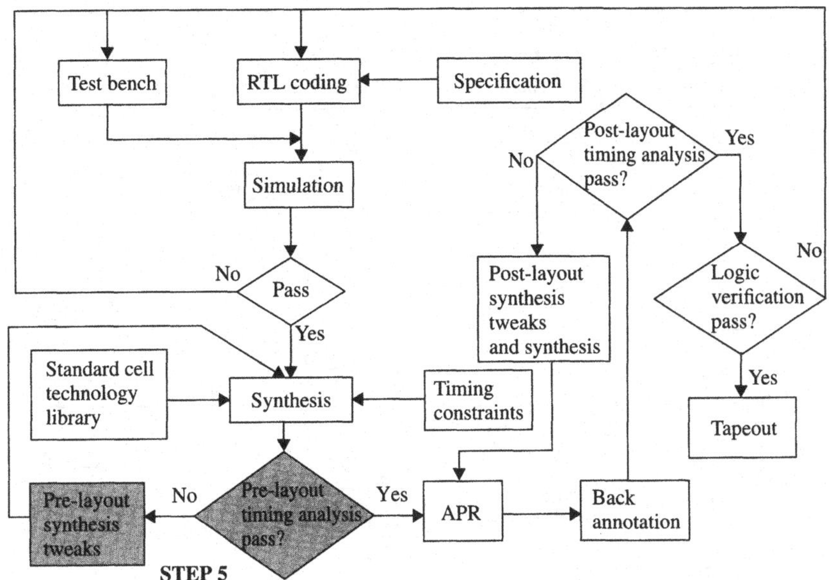

FIGURE 2.7. Diagram indicating Step 5 of an ASIC design flow: pre-layout timing analysis.

A common fix for hold violation is to add delay cells into the path that is failing hold time check. A common fix for setup violation is to reduce the overall delay of the path that failed the setup timing check.

These synthesis tweaks are used to resynthesize the design. Another pre-layout timing analysis is performed.

Step 5 in the ASIC flow sometimes varies depending on the design project. Some design projects will proceed to Step 6, although having timing failures in pre-layout timing analysis. The reason for this is because it is pre-layout timing analysis. The interconnect parasitics that are used for timing analysis are estimations and may not be accurate.

A more common method used in Step 5 is to fix timing failures that are above certain values. The designer can set a value of *x* nanoseconds allowed timing violation. The path that fails more than *x* nanoseconds is fixed. The path that fails less than *x* nanoseconds is not fixed. Again, this can be attributed to the fact that the parasitics used in the timing analysis are not accurate, because no back annotated information is used during this step (pre-layout timing analysis).

2.6 APR

Once pre-layout timing analysis of the synthesized database is completed, the synthesized database together with the timing information from synthesis is

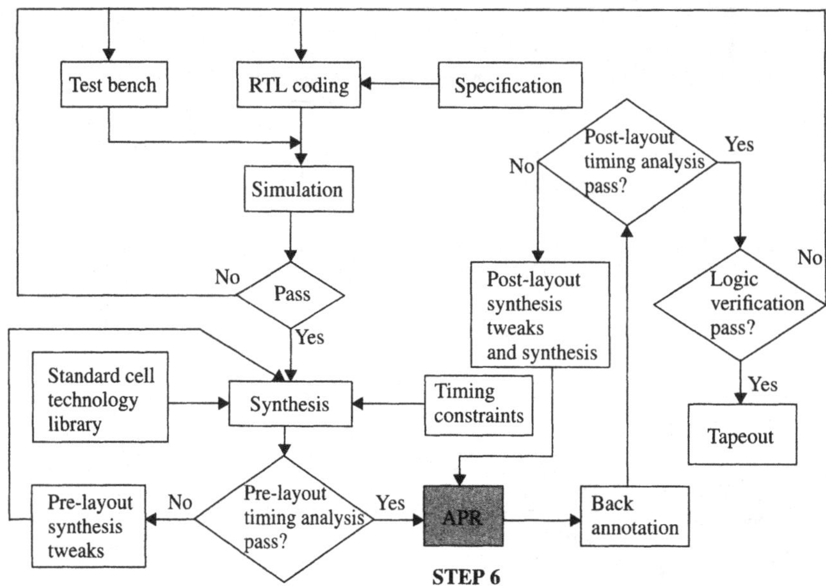

FIGURE 2.8. Diagram indicating Step 6 of an ASIC design flow: APR.

used for APR (Fig. 2.8). In this step, synthesized logic gates are placed and routed. The process of this placement and routing has some degree of flexibility whereby the designer can place the logic gates of each submodule according to a predefined floor plan.

Most designs have critical paths that are tight in terms of timing. These paths can be specified by the designer as high-priority paths. The APR tool will route these high-priority paths first before routing other paths to allow for the most optimal routing.

APR is also the step involved in clock tree synthesis. Most APR tools can handle routing of clock tree with built-in special algorithms. This is an especially important portion of the APR flow because it is critical that the clock tree be "routed" correctly with an acceptable clock skew. Most APR tools allow a designer to specify a required clock skew and buffers up each branch on the clock tree to the desired clock skew.

2.7 BACK ANNOTATION

Back annotation is the step in the ASIC design flow where the RC parasitics in layout is extracted (Fig. 2.9). The path delay is calculated from these RC parasitics. For deep submicron design, these parasitics can cause a significant increase in path delay. Long routing lines can significantly increase the interconnect delay for a path. This could potentially cause paths that are previously

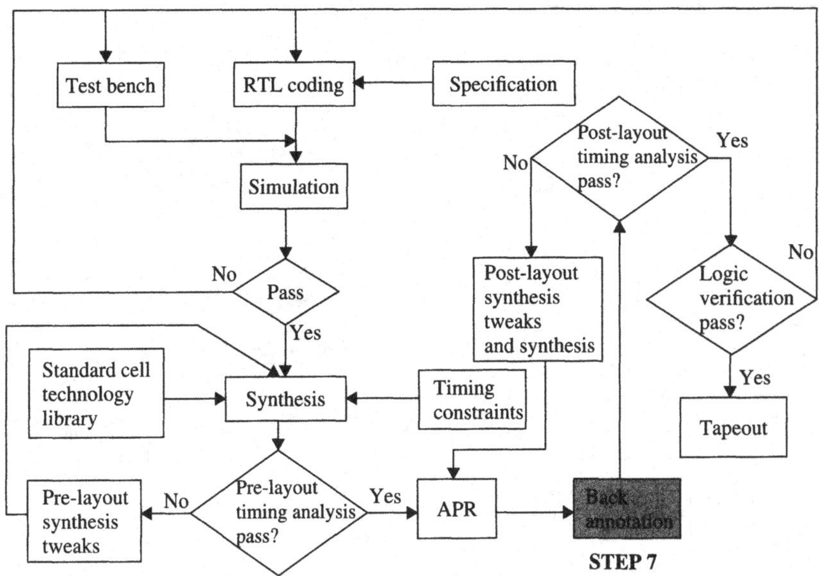

FIGURE 2.9. Diagram indicating Step 7 of an ASIC design flow: back annotation.

(in pre-layout) not critical in timing to be timing critical. It could also cause paths that are meeting the timing requirements to now become critical paths that no longer meet the timing requirements.

Back annotation is an important step that bridges the differences between synthesis and physical layout. During synthesis, design constraints are used by the synthesis tool to generate the logic that is required. However, these design constraints are only an estimation of constraints that apply to each module. The real physical constraints caused by the RC parasitics may or may not reflect the estimated constraints accurately. More likely than not, the estimations are not accurate. As a result, these will cause differences between synthesis and physical layout. Back annotation is the step that bridges them.

2.8 POST-LAYOUT TIMING ANALYSIS

Post-layout timing analysis is an important step in ASIC design flow that allows real timing violations such as hold and setup, to be caught (Fig. 2.10). This step is similar to pre-layout timing analysis, but it includes physical layout information.

In this step, the net interconnect delay information from back annotation is fed into a timing analysis tool to perform post-layout timing analysis. Any setup violations need to be fixed by optimizing the paths that fail the setup

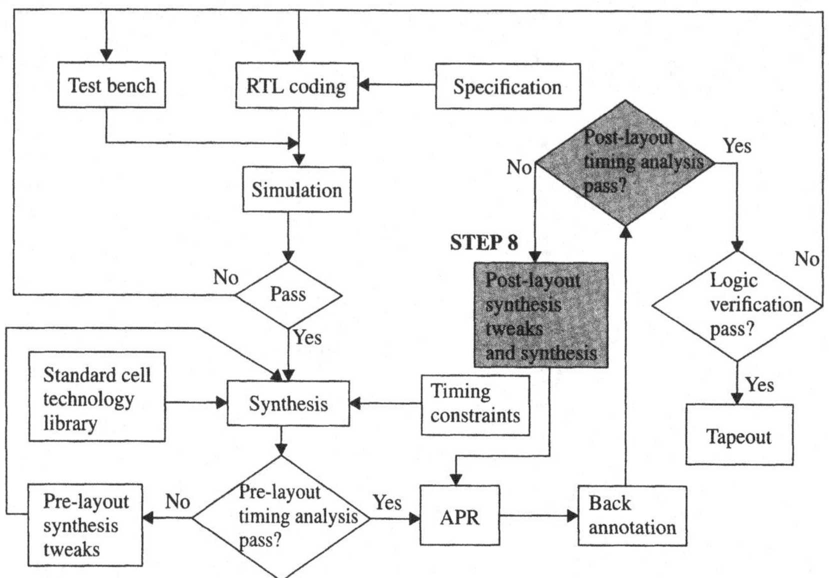

FIGURE 2.10. Diagram indicating Step 8 of an ASIC design flow: post-layout timing analysis.

violations to reduce the path delay. Any hold violation is fixed by adding buffers to the path to increase the path delay.

Post-layout synthesis tweaks are used to make these timing fixes during resynthesis. This allows logic optimization of those failing paths.

When post-layout synthesis is completed, APR, back annotation, and timing analysis are performed again. This will occur in a loop until all the timing violations are fixed. When there are no longer timing violations in the layout database, the design is ready for logic verification.

Note: Post-layout timing analysis is the same as pre-layout timing analysis, except that in post-layout timing analysis, accurate net delay information from physical layout (net delay information for the design is obtained from the extracted layout parasitics) is used. In pre-layout timing analysis, net delay information is estimated.

2.9 LOGIC VERIFICATION

When post-layout timing analysis is completed, the next step is logic verification (Fig. 2.11). This step acts as a final sanity check to ensure the design has the correct functionality. In this step, the design is resimulated using the existing test benches used in Step 3 but with additional timing information obtained from layout.

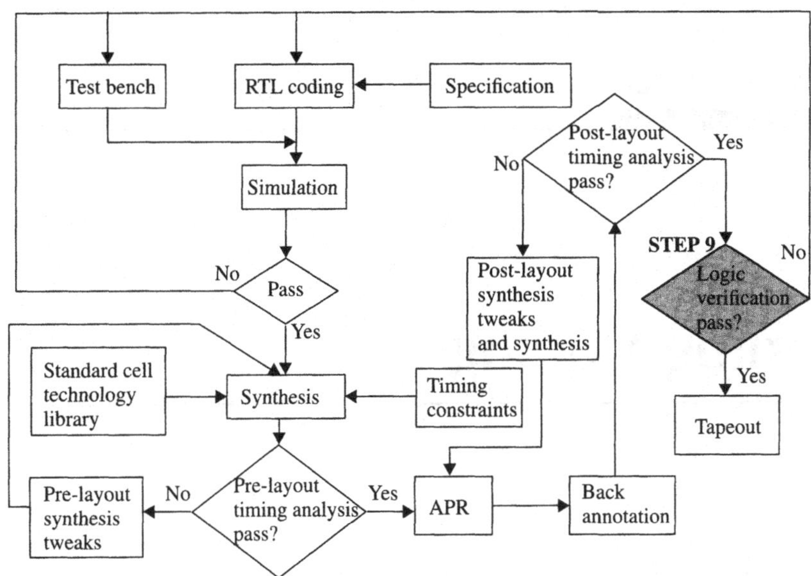

FIGURE 2.11. Diagram indicating Step 9 of an ASIC design flow: logic verification.

Although the design has been verified in Step 3, the design may have failures in Step 9. The failures may be caused by timing glitches or race conditions due to layout parastics. If there are failures, the designer has to fix these failures by either moving back to Step 2 (RTL coding) or Step 8 (post-layout synthesis tweaks).

When the design has finally passed logic verification, it proceeds to tapeout.

CHAPTER THREE

Verilog Coding

3.1 INTRODUCTION TO BASIC VERILOG CONCEPTS

Verilog is a widely used hardware description language (HDL) for design of digital circuits. It can also be used for modeling analog circuits. Whichever it is used for, the basic concept of Verilog remains the same.

When a designer writes Verilog code, it is important to know some of the basic symbols used in Verilog.

3.1.1 Verilog Syntax

Verilog is a HDL that allows a designer to describe a hardware design. As with all languages, there is a required syntax when writing Verilog code.

All Verilog syntax begins with a module declaration. A module is essentially a "box" or "unit" containing the design. The module declaration must include the module's interface ports:

```
module design_module_name (interface_port_list);
```

whereby **design_module_name** is the name of the module and **interface_port_list** is a list of all the input, output, and inout ports to the module. Each port is separated by a comma (,).

The type of interface port is declared. It can be input, output, or inout for bidirectional ports:

Verilog Coding for Logic Synthesis, edited by Weng Fook Lee
ISBN 0-471-429767

```
module DUT (A, B, C, D, E);
input A, B, C;
inout D;
output E;
```

If a port has more than one bit, the declaration must use symbol "[" and "]" to denote the bus width.

```
module DUT (A, B, C, D, E);
input [3:0] A, B;
input C;
inout [7:0] D;
output E;
```

3.1.2 Comments

When writing HDL code for a design, it is a good writing habit for the designer to use comments. It is a good method of indicating to a reader what the code is being written for. It also serves as a good form of documentation that can easily be referred to in the future.

Verilog allows single- or multiple-line comments. The single-line comments use the symbol //, whereas the multiple-line comments begin with the symbol /* and end with the symbol */. For example,

```
// this is a single-line comment in Verilog

/* this is a multiple-line comment in Verilog. Notice
that it begins with a certain symbol and ends with
a certain symbol */
```

3.1.3 Numbers

Verilog allows a wide range of numbers to be used while coding. A designer may choose to use real numbers, integer numbers, base numbers, time domain numbers, signed numbers, and unsigned numbers.

1. *Real numbers* can be declared in Verilog by using the keyword ***real***. It allows numbers to be declared in either a decimal or scientific format. Real numbers can also be declared as negative value.

```
module real_example();

real a,b,c;

initial
begin
```

```
a = 3.141593;
b = 3141e-3;          ──────────►  Real numbers
c = -1.11;
end

endmodule
```

2. *Integer numbers* are declared in Verilog by using the keyword ***integer***. It can also consist of number that are negative in value.

```
module example ();

integer i,j,k;

initial
begin
      i = 150;
      j = -150;         ──────────►  Integer numbers
      k = -32;
end

endmodule
```

3. *Base numbers* are basically integer numbers but declared using a certain base value. They can be octal, hexadecimal, decimal, or binary. Base numbers are declared in the following format:

```
<integer_name> = <bit_size>'<base_namber><value>;
```

Whereby:

- <integer_name> is the name of the integer.
- <bit_size> is the number of binary bits that are representing the integer.
- <base_number> is the base number. It can be in *o* (octal), *h* (hexadecimal), *d* (decimal), or *b* (binary) format.
- <value> is the value of the integer.

```
module example ();

integer i,j,k,l;

initial
begin
   i = 5'b10111; // this is a binary number
   j = 5'o24; // this is an octal number
   k = 8'ha9; // this is a hex number
   l = 5'd24; // this is a decimal number
end

endmodule
```

4. *Time domain numbers*. Time in Verilog simulation is declared with the keyword ***time***. The unit for time is declared using Verilog compiler directives of timescale. The declaration of timescale must be in the following format:

```
timescale <reference_time>/<precision>;
```

whereby <reference_time> and <precision> must be in integer values of either 1, 10, or 100. However, time units are allowed to be specified with these integers to define time in fs (femtosecond), ps (picosecond), ns (nanosecond), µs (microsecond), ms (millisecond), and s (second).

```
module example ();

`timescale 100 µs / 1 ns; // this is for
// reference of 100 µs and
  // precision of 1 ns

time t;

initial
begin
  t = $time; // $time is a Verilog system function
             // that gets the current simulation time
end

endmodule
```

3.1.4 Verilog Data Type

Verilog allows two data types, ***reg*** and net. ***Reg*** (short for register) is a storage element. It allows values to be stored in data type. These values are kept in the data type until they are replaced by other values. ***Reg*** can only be used in an ***always*** statement or ***initial*** statement.

The most common net type used in verilog is ***wire***. It is commonly used to represent net connection. It is analogous to a physical wire in hardware. Therefore, the value that is on a ***wire*** is continuously updated.

During simulation, if no values are assigned to any identifiers declared under ***reg*** type, the default value of the identifier is an unknown or *X*. Similarly, if no values are assigned to any identifiers declared under ***wire*** type, the default value of the identifier is a tri-state or *Z*.

Example 3.1 shows a simple method of using ***wire*** while Example 3.2 is similar to Example 3.1, except that it is a 4-bit bus declaration.

Example 3.1 Verilog Code Using Wire Declaration

```
module example (inputA, inputB, inputC, outputA);
```

```
input inputA, inputB, inputC;

output outputA;

wire temp;

assign temp = inputB | inputC;

assign outputA = inputA & temp;

endmodule
```

Example 3.2 Verilog Code Using Wire Declaration for a 4-bit Bus

```
module example (inputA, inputB, inputC, outputA);

input [3:0] inputA, inputB, inputC;
output [3:0] outputA;

wire [3:0] temp;

assign temp = inputB | inputC;

assign outputA = inputA & temp;

endmodule
```

Example 3.3 shows a common method of using ***reg*** whereas Example 3.4 is similar to Example 3.3 except that it is an 8-bit bus declaration.

Example 3.3 Verilog Code Using Reg Declaration

```
module example (inputA, inputB, inputC, outputA);

input inputA, inputB, inputC;
output outputA;

reg outputA, temp;

always @ (inputA or inputB or inputC)
begin
     if (inputA)
          temp = 1'b0;
```

```
    else
        begin
            if (inputB & inputC)
                temp = 1'b1;
            else
                temp = 1'b0;
        end
end

// more source code

always @ (temp or inputC or inputA)
begin
    if (temp)
        outputA = inputC;
    else
        outputA = inputA;
end

endmodule
```

Example 3.4 Verilog Code Using Reg Declaration for an 8-bit Bus

```
module example (inputA, inputB, inputC, outputA);

input [7:0] inputA, inputB, inputC;
output [7:0] outputA;

reg [7:0] outputA, temp;

always @ (inputA or inputB or inputC)
begin
    if (inputA)
        temp = 8'b11110000;
    else
        begin
            if (inputB & inputC)
                temp = 8'b10100101;
            else
                temp = 8'b01011010;
        end
end

// more source code
```

```
always @ (temp or inputC or inputA)
begin
     if (temp == 8'b10100101)
          outputA = inputC;
     else
          outputA = inputA;
end

endmodule
```

Apart from ***wire*** and ***reg***, there are 10 other net types that can be used in Verilog.

1. ***supply1***, as the name implies, is used on nets that are connected to a VCC supply. It is declared in Verilog using the keyword ***supply1***, supplyl VCC;
2. ***supply0***, as the name implies, is used on nets that are connected to ground. It is declared in Verilog using the keyword ***supply0***, *supply0 VSS*;
3. ***tri*** is a net type that is used to declare a net that has more than one driver driving it. Example 3.5 shows a Verilog code that has a net ***temp*** that is driven by more than one driver.

Example 3.5 Verilog Code Showing a Tri Declaration

```
module example (inputA, inputB, inputC, outputA);

input inputA, inputB,
inputC; output outputA;

tri temp;

assign temp = inputA
& ~inputB;

// Verilog source code

assign temp = inputA
| ~inputB;

assign outputA = temp
& inputC;

endmodule
```

```
Net type tri is
synthesizable. However, it
is not advisable to use
net type tri when writing
synthesizable Verilog. If
a node is to be driven by
multiple drivers, that
node should be driven only
by tristate drivers. The
example shown here using
net type tri on node temp
with multiple assign
statements driving it is
not a good coding method
in synthesis.
```

4. ***trior*** is also used for a net that has more than one driver driving it. However, it is different from ***tri*** as ***trior*** is a wired-OR type of net connection. This means that if any of the drivers that drive the ***trior*** net is at a logical 1, the ***trior*** net will be at logical 1 as well. ***trior*** is not synthesizable and cannot be used when coding for synthesis.
5. ***triand*** is also used for a net that has more than one driver driving it. However, it is different from ***tri***, as ***triand*** is a wired-AND type of net connection. This means that if any of the drivers that drives the ***triand*** net is at a logical 0, the ***triand*** net will be at logical 0 as well. ***triand*** is not synthesizable and cannot be used when coding for synthesis.
6. ***trireg*** is also used for a net that has more than one driver driving it. However, it is different from ***tri***, as ***trireg*** nets are capacitive nets. This means that ***trireg*** nets have the ability to store a value. If the drivers that drive the ***trireg*** nets are at *Z* state or high impedance state, the ***trireg*** net will maintain its value on the net. ***trireg*** is not synthesizable and cannot be used when coding for synthesis.
7. ***tri1*** is also used for a net that has more than one driver driving it. However, it is different from ***tri***, as ***tri1*** net is at a logical 1 state if the drivers that drive the ***tri1*** net is at *Z* state or high impedance. ***tri1*** is not synthesizable and cannot be used when coding for synthesis.
8. ***tri0*** is also used for a net that has more than one driver driving it. However, it is different from ***tri***, as ***tri0*** net is at a logical 0 state if the drivers that drive the ***tri0*** net is at *Z* state or high impedance. ***tri0*** is not synthesizable and cannot be used when coding for synthesis.
9. ***wand*** is used for nets with a wired-AND configuration, whereby if any of the drivers driving the ***wand*** net is at logical 0, the ***wand*** net will be at logical 0. ***wand*** net is synthesizable.
10. ***wor*** is used for nets with a wired-OR configuration, whereby if any of the drivers driving the ***wor*** net is at logical 1, the ***wor*** net will be a logical 1. ***wor*** net is synthesizable.

Note: When coding for synthesis, the most commonly used type of declaration for net is ***wire****. Net types* ***tri, wand****, and* ***wor*** *are synthesizable but not advisable for use in synthesizable Verilog code. Net types* ***trior, triand, trireg, tri1****, and* ***tri0*** *are not synthesizable.*

In Verilog, each net or ***reg*** can have one of four values:

1 – represents logical 1
0 – represents logical 0
X – represents don't care state
Z – represents high impedance

For nets with conditions of multiple drivers driving them, each driver having the possibility of driving any one of four values stated, what value would the net be at?

Assume net C is driven by two drivers, A and B. Both drivers can each drive any one of the four values of *1*, *0*, *X*, or *Z*, thereby allowing a possible combination of 16 conditions on the drivers. The final value on net C due to the 16 different driving conditions would depend on the net type that has been declared on net C.

1. ***tri*** Referring to Table 3.1, if driver A is driving a value of logic 0 and driver B is driving a value of logic 1, net C, which is declared as a ***tri*** net type, will have a value of *X*. *Note: Net type **wire** has the same value as the net type **tri** for multiple drivers driving the net.*
2. ***trior*** Referring to Table 3.2, if driver A or B is driving a value of logic 1, net C, which is declared as a ***trior*** net type, will have a value of 1.
3. ***triand*** Referring to Table 3.3, if driver A or B is driving a value of logic 0, net C, which is declared as a ***triand*** net type, will have a value of 0.
4. ***trireg*** Referring to Table 3.4, if drivers A and B are tristated, net C, which is declared as a ***trireg*** net type, will hold its previous value.
5. ***tri1*** Referring to Table 3.5, if drivers A and B are tristated, net C which is declared as a ***tri1*** net type, will have a value of 1.

TABLE 3.1. Table indicating value on net C (net type *tri*) for different net values on drivers A and B

tri		*A*			
		1	0	Z	X
	1	*1*	*X*	*1*	*X*
	0	*X*	*0*	*0*	*X*
B	Z	*1*	*0*	*Z*	*X*
	X	*X*	*X*	*X*	*X*

TABLE 3.2. Table indicating value on net C (net type *trior*) for different net values on drivers A and B

Trior		*A*			
		1	0	Z	X
	1	*1*	*1*	*1*	*1*
	0	*1*	*0*	*0*	*X*
B	Z	*1*	*0*	*Z*	*X*
	X	*1*	*X*	*X*	*X*

TABLE 3.3. Table indicating value on net C (net type *triand*) for different net values on drivers A and B

Triand		A			
		1	0	Z	X
	1	*1*	*0*	*1*	*X*
	0	*0*	*0*	*0*	*0*
B	Z	*1*	*0*	*Z*	*X*
	X	*X*	*0*	*X*	*X*

TABLE 3.4. Table indicating value on net C (net type *trireg*) for different net values on drivers A and B

Trireg		A			
		1	0	Z	X
	1	*1*	*X*	*1*	*X*
	0	*X*	*0*	*0*	*X*
B	Z	*1*	*0*	*Previous value*	*X*
	X	*X*	*X*	*X*	*X*

TABLE 3.5. Table indicating value on net C (net type *tri1*) for different net values on drivers A and B

Tril		A			
		1	0	Z	X
	1	*1*	*X*	*1*	*X*
	0	*X*	*0*	*0*	*X*
B	Z	*1*	*0*	*1*	*X*
	X	*X*	*X*	*X*	*X*

6. ***tri0*** Referring to Table 3.6, if drivers A and B are tristated, net C, which is declared as a ***tri0*** net type, will have a value of 0.
7. ***wand*** Referring to Table 3.7, if drivers A or B are driving a value of logic 0, net C, which is declared as a ***wand*** net type, will have a value of 0.
8. ***wor*** Referring to Table 3.8, if drivers A or B are driving a value of logic 1, net C, which is declared as a ***wor*** net type, will have a value of 1.

3.1.5 Signal Strength

Section 3.1.3 has discussed in detail on the different types of net declarations as well as the usage of ***reg*** in Verilog. Each net type or ***reg*** can have the value

TABLE 3.6. Table indicating value on net C (net type *tri0*) for different net values on drivers A and B

Tri0		*A*			
		1	0	Z	X
	1	*1*	*X*	*1*	*X*
	0	*X*	*0*	*0*	*X*
B	Z	*1*	*0*	*0*	*X*
	X	*X*	*X*	*X*	*X*

TABLE 3.7. Table indicating value on net C (net type *wand*) for different net values on drivers A and B

Wand		*A*			
		1	0	Z	X
	1	*1*	*0*	*1*	*X*
	0	*0*	*0*	*0*	*0*
B	Z	*1*	*0*	*Z*	*X*
	X	*X*	*0*	*X*	*X*

TABLE 3.8. Table indicating value on net C (net type *wor*) for different net values on drivers A and B

Wor		*A*			
		1	0	Z	X
	1	*1*	*1*	*1*	*1*
	0	*1*	*0*	*0*	*X*
B	Z	*1*	*0*	*Z*	*X*
	X	*1*	*X*	*X*	*X*

of 0, 1, X, or Z. The values of a net or ***reg***, although limited to only these four values, can have eight different strengths (Table 3.9). The level of strength of a ***wire*** or ***reg*** is often used to resolve a situation when contention occurs.

Note: When coding for synthesis, strength levels are seldom used. This is because strength levels are used to resolve contentions within a logic circuit. However, when coding for synthesis, it is a good coding practice to ensure that the Verilog code does not have contention. An example of a design that has contention is discussed in detail in Chapter 5.

Example 3.6 shows a simple Verilog code that assigns strength values to the output of a design.

TABLE 3.9. Table showing different strength levels

Strength Type	*Level*
Supply	Strongest
Strong	↑
Pull	
Large	
Weak	
Medium	
Small	
High impedance	Weakest

Example 3.6 Verilog Code Using Strength Assignment

```
module example (inputA, inputB, inputC, outputA,
outputB);

input inputA, inputB, inputC;
output outputA, outputB;

wire outputA, outputB;

and (strong1, weak0) and_gate_instance (outputA,
inputA, inputB);
or (weak1, weak0) or_gate_instance (outputB, inputB,
inputC);

endmodule
```

```
During synthesis, synthesis tools would
ignore the strength assignments. This
example would synthesize to an AND gate
and an OR gate.
```

3.2 VERILOG GATE-LEVEL PRIMITIVES

Verilog allows coding to include gate-level primitives that can be instantiated in Verilog code. These primitives are built-in as part of Verilog coding and do not require any special setup.

Some of these primitives are synthesizable, whereas others are not. The following lists the available gate-level primitives that can be used in Verilog coding:

1. ***pmos*** This primitive is used to represent a pmos transistor. It has two inputs and one output and can be modeled as follows:

```
pmos pmos_instance (Output_signal, Input_signal,
Gate_signal);
```

whereby ***pmos_instance*** is the name of the instance of the instantiated ***pmos*** transistor, ***Output_signal*** is the name of the net that is connected to the output of the ***pmos*** transistor, ***Input_signal*** is the name of the net that is connected to the input of the ***pmos*** transistor, and ***Gate_signal*** is the name of the net that is connected to the gate of the ***pmos*** transistor (Table 3.10).

2. ***nmos*** This primitive is used to represent an ***nmos*** transistor. It has two inputs and one output and can be modeled as follows:

```
nmos nmos_instance (Output_signal, Input_signal,
Gate_signal);
```

whereby ***nmos_instance*** is the name of the instance of the instantiated ***nmos*** transistor, ***Output_signal*** is the name of the net that is connected to the output of the ***nmos*** transistor, ***Input_signal*** is the name of the net that is connected to the input of the ***nmos*** transistor, and ***Gate_signal*** is the name of the net that is connected to the gate of the ***nmos*** transistor (Table 3.11).

3. ***cmos*** This primitive is used to represent a cmos passgate. It has three inputs and one output and can be modeled as follows:

```
cmos cmos_instance (Output_signal, Input_signal,
NGate_signal, PGate_signal);
```

TABLE 3.10. Truth table for *pmos* transistor primitive

Input_signal	Gate_signal	Output_signal
0	0	0
0	1	Z
1	0	1
1	1	Z

TABLE 3.11. Truth table for *nmos* transistor primitive

Input_signal	Gate_signal	Output_signal
0	0	Z
0	1	0
1	0	Z
1	1	1

whereby ***cmos_instance*** is the name of the instance of the instantiated ***cmos*** passgate, ***Output_signal*** is the name of the net that is connected to the output of the ***cmos*** passgate, ***Input_signal*** is the name of the net that is connected to the input of the ***cmos*** passgate, ***NGate_signal*** is the name of the net that is connected to the N side control gate of the ***cmos*** passgate, and ***PGate_signal*** is the name of the net that is connected to the P side control gate of the ***cmos*** passgate (Table 3.12).

4. ***rpmos*** This primitive behaves the same as ***pmos*** except that ***rpmos*** is more resistive in nature compared with ***pmos***. This would result in the output of this primitive having a reduction in strength compared with the output of the ***pmos*** primitive.
5. ***rnmos*** This primitive behaves the same as the primitive ***nmos*** except that ***rnmos*** is more resistive in nature compared with ***nmos***. This would result in the output of the primitive having a reduction in strength compared with the output of the ***nmos*** primitive.
6. ***rcmos*** This primitive behaves the same as the primitive ***cmos*** except that ***rcmos*** is more resistive in nature compared with ***cmos***. This would result in the output of the primitive having a reduction in strength compared with the output of the ***cmos*** primitive.
7. ***pullup*** This primitive, as its name implies, is used to represent a pullup node. It can be modeled as follows:

```
pullup pullup_instance (signal_name);
```

   whereby ***pullup_instance*** is the name of the instance of the pullup and ***signal_name*** is the name of the signal that is being "pulled up."
8. ***pulldown*** This primitive, as its name implies, is used to represent a pulldown node. It can be modeled as follows:

```
pulldown pulldown_instance (signal_name);
```

   whereby ***pulldown_instance*** is the name of the instance of the pulldown and ***signal_name*** is the name of the signal that is being "pulled down."

TABLE 3.12. Truth table for *cmos* passgate primitive

Input_signal	NGate_signal	PGate_signal	Output_signal
0	0	0	0
0	0	1	Z
0	1	0	0
0	1	1	0
1	0	0	1
1	0	1	Z
1	1	0	1
1	1	1	1

9. ***tran*** This primitive is used to represent a bi-directional switch that allows data to flow both ways between two nets. It can be modeled as follows:

   ```
   tran tran_instance (netA, netB);
   ```

   whereby ***tran_instance*** is the name of the instance, and ***netA*** and ***netB*** are the names of two nets on which data can flow between them.
10. ***rtran*** This primitive behaves the same as primitive ***tran*** except that ***rtran*** is more resistive in nature compared with ***tran***. This would result in the output of the primitive having a reduction in strength compared with the output of the ***tran*** primitive.
11. ***tranif0*** This primitive behaves the same as primitive ***tran*** except that ***tranif0*** only allows data flow between two nets if a gate control signal is at a logic 0. Otherwise, data flow is disabled. The ***tranif0*** primitive can be modeled as follows:

    ```
    tranif0 tranif0_instance (netA, netB,
    Gate_control);
    ```

    whereby ***tranif0_instance*** is the name of the instance, ***netA*** and ***netB*** are the names of the two nets that have the data flow between each of them, and ***Gate_control*** is the name of the signal that will only allow data flow when it is at logical 0.
12. ***tranif1*** This primitive behaves the same as primitive ***tran*** except that ***tranif1*** only allows data flow between two nets if a gate control signal is at logic 1. Otherwise, data flow is disabled. The ***tranif1*** primitive can be modeled as follows:

    ```
    tranif1 tranif1_instance (netA, netB,
    Gate_control);
    ```

    whereby ***tranif1_instance*** is the name of the instance, ***netA*** and ***netB*** are the names of the two nets that have data flow between each of them, and ***Gate_control*** is the name of the signal that will only allow data flow when it is at logical 1.
13. ***rtranif0*** This primitive behaves the same as primitive ***tranif0*** except that ***rtranif0*** is more resistive in nature compared with ***tranif0***. This would result in the output of the primitive having a reduction in strength compared with the output of the ***tranif0*** primitive.
14. ***rtranif1*** This primitive behaves the same as primitive ***tranif1*** except that ***rtranif1*** is more resistive in nature compared with ***tranif1***. This would result in the output of the primitive having a reduction in strength compared with the output of the ***tranif1*** primitive.
15. ***notif0*** This primitive is used to represent a tri-state inverter. It has two inputs and one output and can be modeled as follows:

```
notif0 notif0_instance (Output_signal,
Input_signal, Control_signal);
```

whereby ***notif0_instance*** is the name of the instance of the instantiated ***notif0*** tri-state inverter, ***Output_signal*** is the name of the net that is connected to the output of the ***notif0*** tri-state inverter, ***Input_signal*** is the name of the net that is connected to the input of the ***notif0*** tri-state inverter, and ***Control_signal*** is the name of the net that is connected to the select input of the ***notif0*** tri-state inverter (Table 3.13).

16. ***notif1*** This primitive is used to represent a tri-state inverter. It has two inputs and one output and can be modeled as follows:

```
notif1 notif1_instance (Output_signal,
Input_signal, Control_signal);
```

whereby ***notif1_instance*** is the name of the instance of the instantiated ***notif1*** tri-state inverter, ***Output_signal*** is the name of the net that is connected to the output of the ***notif1*** tri-state inverter, ***Input_signal*** is the name of the net that is connected to the input of the ***notif1*** tri-state inverter, and ***Control_signal*** is the name of the net that is connected to the select input of the ***notif1*** tri-state inverter (Table 3.14).

17. ***bufif0*** This primitive is used to represent a tri-state buffer. It has two inputs and one output and can be modeled as follows:

```
bufif0 bufif0_instance (Output_signal, Input_signal,
Control_signal);
```

TABLE 3.13. Truth table for notif0 tri-state inverter primitive

Input_signal	Control_signal	Output_signal
0	0	1
0	1	Z
1	0	0
1	1	Z

TABLE 3.14. Truth table for notif1 tri-state inverter primitive

Input_signal	Control_signal	Output_signal
0	0	Z
0	1	1
1	0	Z
1	1	0

whereby ***bufif0_instance*** is the name of the instance of the instantiated ***bufif0*** tri-state buffer, ***Output_signal*** is the name of the net that is connected to the output of the ***bufif0*** tri-state buffer, ***Input_signal*** is the name of the net that is connected to the input of the ***bufif0*** tri-state buffer, and ***Control_signal*** is the name of the net that is connected to the select input of the ***bufif0*** tri-state buffer (Table 3.15).

18. ***bufif1*** This primitive is used to represent a tri-state buffer. It has two inputs and one output and can be modeled as follows:

```
bufif1 bufif1_instance (Output_signal, Input_signal,
Control_signal);
```

whereby ***bufif1_instance*** is the name of the instance of the instantiated ***bufif1*** tri-state buffer, ***Output_signal*** is the name of the net that is connected to the output of the ***bufif1*** tri-state buffer, ***Input_signal*** is the name of the net that is connected to the input of the ***bufif1*** tri-state buffer, and ***Control_signal*** is the name of the net that is connected to the select input of the ***bufif1*** tri-state buffer (Table 3.16).

19. ***buf*** This primitive is used to represent a buffer. It has one input and one or more outputs and can be modeled as follows:

```
buf buf_instance (Output_signal, Input_signal);
```

whereby ***buf_instance*** is the name of the instance of the instantiated ***buf*** buffer, ***Output_signal*** is the name of the net that is connected to the output of the ***buf*** buffer, and ***Input_signal*** is the name of the net

TABLE 3.15. Truth table for bufif0 tri-state buffer primitive

Input_signal	Control_signal	Output_signal
0	0	0
0	1	Z
1	0	1
1	1	Z

TABLE 3.16. Truth table for bufif1 tri-state buffer primitive

Input_signal	Control_signal	Output_signal
0	0	Z
0	1	0
1	0	Z
1	1	1

that is connected to the input of the ***buf*** buffer. For cases where there is more than one output, it can be modeled as follows:

```
buf buf_instance (Output_signal1, Output_signal2,
Output_signal3, Output_signal4, Input_signal);
```

whereby ***Output_signal1***, ***Output_signal2***, ***Output_signal3***, and ***Output_signal4*** are all the outputs of the buffer.

20. ***not*** This primitive is used to represent an inverter. It has one input and one or more outputs and can be modeled as follows:

    ```
    not not_instance (Output_signal, Input_signal);
    ```

    whereby ***not_instance*** is the name of the instance of the instantiated ***not*** inverter, ***Output_signal*** is the name of the net that is connected to the output of the ***not*** gate, and ***Input_signal*** is the name of the net that is connected to the input of the ***not*** gate. For cases where there is more than one output, it can be modeled as follows:

    ```
    not not_instance (Output_signal1, Output_signal2,
    Output_signal3, Output_signal4, Input_signal);
    ```

    whereby ***Output_signal1***, ***Output_signal2***, ***Output_signal3***, and ***Output_signal4*** are the output of the inverter.

21. ***and*** This primitive is used to represent an AND gate. It can have two or more inputs and one output. It can be modeled as follows:

    ```
    and and_instance (Output_signal, Input_signal1,
    Input_signal2);
    ```

    whereby ***and_instance*** is the name of the instance of the AND gate, ***Output_signal*** is the name of the net connected to the output of the AND gate, and ***Input_signal1*** and ***Input_signal2*** are the names of the nets that are connected to the inputs of the AND gate.

22. ***nand*** This primitive is used to represent a NAND gate. It can have two or more inputs and one output. It can be modeled as follows:

    ```
    nand nand_instance (Output_signal, Input_signal1,
    Input_signal2);
    ```

    whereby ***nand_instance*** is the name of the instance of the NAND gate, ***Output_signal*** is the name of the net connected to the output of the NAND gate, and ***Input_signal1*** and ***Input_signal2*** are the names of the nets that are connected to the inputs of the NAND gate.

23. ***nor*** This primitive is used to represent a NOR gate. It can have two or more inputs and one output. It can be modeled as follows:

    ```
    nor nor_instance (Output_signal, Input_signal1,
    Input_signal2);
    ```

whereby ***nor_instance*** is the name of the instance of the NOR gate, ***Output_signal*** is the name of the net connected to the output of the NOR gate, and ***Input_signal1*** and ***Input_signal2*** are the names of the nets that are connected to the inputs of the NOR gate.

24. ***or*** This primitive is used to represent an OR gate. It can have two or more inputs and one output. It can be modeled as follows:

```
or or_instance (Output_signal, Input_signal1,
Input_signal2);
```

whereby ***or_instance*** is the name of the instance of the OR gate, ***Output_signal*** is the name of the net connected to the output of the OR gate, and ***Input_signal1*** and ***Input_signal2*** are the names of the nets that are connected to the inputs of the OR gate.

25. ***xor*** This primitive is used to represent a XOR gate. It can have two or more inputs and one output. It can be modeled as follows:

```
xor xor_instance (Output_signal, Input_signal1,
Input_signal2);
```

whereby ***xor_instance*** is the name of the instance of the XOR gate, ***Output_signal*** is the name of the net connected to the output of the XOR gate, and ***Input_signal1*** and ***Input_signal2*** are the names of the nets that are connected to the inputs of the XOR gate.

26. ***xnor*** This primitive is used to represent an XNOR gate. It can have two or more inputs and one output. It can be modeled as follows:

```
xnor xnor_instance (Output_signal, Input_signal1,
Input_signal2);
```

whereby ***xnor_instance*** is the name of the instance of the XNOR gate, ***Output_signal*** is the name of the net connected to the output of the XNOR gate, and ***Input_signal1*** and ***Input_signal2*** are the names of the nets that are connected to the inputs of the XNOR gate.

Note: When coding for synthesis, the designer must be careful about which primitives are being used. Not all the gate primitives are synthesizable. Gate primitives that can be used in coding for synthesis are ***or, and, xor, nor, nand, xnor, not, buf, bufif0, bufif1, notif0****, and* ***notif1****. The other gate primitives—***tran, tranif0, tranif1, rtran, rtranif0, rtranif1, pullup, pulldown, pmos, nmos, cmos, rpmos, rnmos****, and* ***rcmos****—are not used in writing synthesizable Verilog code.*

3.3 USER-DEFINED PRIMITIVES

Section 3.2 discussed the use of gate-level primitives that are built into Verilog language. Apart from these primitives, a designer can also create their own primitives, which are referred to as user-defined primitives (UDP).

In general, a UDP is a module that is defined and described by the user. This UDP module can be used in Verilog code by instantiating it. There are two types of UDP that a designer may create, combinational UDP and sequential UDP.

3.3.1 Combinational UDP

Combinational UDP describes a module that is combinational in nature. This means that the UDP module consists of combinational logic to create its output.

Example 3.7 shows the syntax defining a combinational UDP.

Example 3.7 Example Showing Syntax Defining a UDP

```
primitive <primitive_UDP_name> (<output_port_list>,
<input_port_list>);
output <output_port_list>;
input <input_port_list>;
table
<Truth_table_format_description_of_combinational_UDP_
functionality>;
endtable
endprimitive
```

whereby:

a. ***<primitive_UDP_name>*** is the name of the UDP primitive being defined.
b. ***<output_port_list>*** is the name of the output port for the UDP primitive. Please note that a UDP primitive can only have one output port, and it can only be one bit wide.
c. ***<input_port_list>*** are the names of all the input ports. Each input port can only be one bit wide.
d. ***<Truth_table_format_description_of_combinational_UDP_functionality>*** is a truth table format description of the functionality of the primitive UDP.

To have an example for a UDP declaration, Table 3.17 is created to describe the functionality of the UDP module.

Example 3.8 Verilog Example for Defining a UDP Primitive and Instantiating the Primitive

```
primitive udp_gate (outputA, inputA, inputB, inputC);
output outputA;
```

TABLE 3.17. Table showing functionality of UDP module UDP_GATE

InputA	InputB	InputC	OutputA
0	0	0	0
0	0	1	0
0	1	0	1
0	1	1	1
1	0	0	0
1	0	1	0
1	1	0	1
1	1	1	1

```
input inputA, inputB, inputC;

table
//   inputA   inputB   inputC   outputA
       0        0        0    :    0;
       0        0        1    :    0;
       0        1        0    :    1;
       0        1        1    :    1;
       1        0        0    :    0;
       1        0        1    :    0;
       1        1        0    :    1;
       1        1        1    :    1;
endtable
endprimitive

// to create a module that instantiates the UDP
// primitive

module example (input1, input2, input3, output1);

input input1, input2, input3;
output output1;

wire output1;

udp_gate udp_gate_inst (output1, input1, input2,
input3);

endmodule
```

Declaration of functionality of UDP.

Instantiation of UDP.

TABLE 3.18. Table showing functionality of UDP module UDP_LATCH

data	clock	Q
0	0	<previous_value>
0	1	0
1	0	<previous_value>
1	1	1

3.3.2 Sequential UDP

Sequential UDP describes a module that is sequential in nature. This means the UDP module consists of storage elements that can store a value.

The syntax for defining a sequential UDP is the same as that for combinational UDP except that it uses ***reg*** declarations.

Table 3.18 shows a truth table for a latch on which a UDP is defined.

Example 3.9 Verilog Example for Defining a Sequential UDP Primitive and Instantiating the Primitive

```
primitive udp_latch (Q, data, clock);
output Q;
input data, clock;

reg Q;

initial
   Q = 0;

table
//   data    clock      Q(current)      Q(next)
       0       0     :      ?       :      -;
       0       1     :      ?       :      0;
       1       0     :      ?       :      -;
       1       1     :      ?       :      1;
endtable
endprimitive

// to create a module that instantiates the sequential
// UDP primitive

module example (qout, indata, inclock);
```

Initialize the latch to logic "0."

The ? represents "don't care" condition while - represents "no change."

TABLE 3.19. Table showing functionality of UDP module UDP_POS_FLOP

data	clock	Q
0	rising edge	0
1	rising edge	1
1	others	<previous_value>
0	others	<previous_value>

```
input indata, inclock;
output qout;

wire qout;

udp_latch udp_latch_inst (qout, indata, inclock);

endmodule
```

Table 3.18 and Example 3.9 are Verilog examples for a level-sensitive sequential element (latch). Table 3.19 and Example 3.10 show a Verilog example for a positive edge-sensitive sequential element (flop).

Example 3.10 Verilog Example for Defining a Positive Edge Sequential UDP Primitive and Instantiating the Primitive

```
primitive udp_pos_flop (Q, data, clock);
output Q;
input data, clock;

reg Q;

initial
   Q = 0;

table
//   data     clock      Q(current)     Q(next)
       0       (01)   :       ?       :     0;
       1       (01)   :       ?       :     1;
// no change in output values
       0       (0x)   :       ?       :     -;
       1       (0x)   :       ?       :     -;
// no change for negedge
       ?       (?0)   :       ?       :     -;
```

```
(01) symbolises a transition
from low to high, (0x)
symbolises a transition from 0
to x, (?0) represents a
transition from any value to a
0.
```

```
// no change for change in data
    (??)     ?    :     ?     :    -;
endtable
endprimitive

// to create a module that instantiates the sequential
// UDP primitive

module example (qout, indata, inclock);

input indata, inclock;
output qout;

wire qout;

udp_pos_flop udp_pos_flop_inst (qout, indata, inclock);

endmodule
```

Note: It is not common practice to code UDP in synthesizable Verilog code. Furthermore, most synthesis tools do not support UDP.

TABLE 3.20. Concurrent and sequential statements

Concurrent	Sequential

wire A, B;
assign A = (input1 | input2) & ~input3;
assign B = (input5 & input6);
	wire A;
	always @ (input1 or input2 or input3 or inputA or input4 or input5 or input6)
	begin
	  if (input1 & ~input2) // Statement 1
	    A = input3;
	  else if (input1 & inputA) // Statement 2
	    A = input4 & input5;
	  else if (input1 & input3) // Statement 3
	    A = input6;
	  else // Statement 4
	    A = 0;
	end

Assignment of A and B occurs concurrently.	Statement 1 is evaluated first, followed by 2, 3 and 4. Only upon completion of statement 1, statement 2 is evaluated. Execution of the evaluation occurs sequentially.

3.4 CONCURRENT AND SEQUENTIAL STATEMENTS

Concurrent and sequential statements are two types of Verilog statements widely used in Verilog coding. Concurrent statements are statements that are executed concurrently. Sequential statements are statements that are executed one after the other (Table 3.20).

CHAPTER FOUR

Coding Style: Best-Known Method for Synthesis

Coding style plays a very important role for an ASIC design flow. "Bad" HDL (either Verilog or VHDL) code does not allow efficient optimization during synthesis. Logic that is generated from synthesis tools depends highly on the code that is written. A badly formed code would generally create bad logic. As the saying goes, "garbage in, garbage out."

There are certain general guidelines to follow when it comes to coding style. By following these guidelines, a constant, good coding style can be attained. By having a good coding style, synthesis results are optimal.

4.1 NAMING CONVENTION

For a design project, a good naming convention is necessary. Naming convention is normally the most overlooked guideline when coding in HDL. Having a well-defined naming convention does not seem to sound important, but not having one can cause a lot of problems in the later stages of design, especially during the fullchip integration. It would be difficult for the designer to connect all the signals between modules of a fullchip if the signal names do not match.

By defining a naming convention, a set of rules is applied when the designer names the ports of a module. If each module in fullchip is based on the same set of naming rules, then it becomes much easier to connect these signals together in the fullchip level.

Verilog Coding for Logic Synthesis, edited by Weng Fook Lee
ISBN 0-471-429767

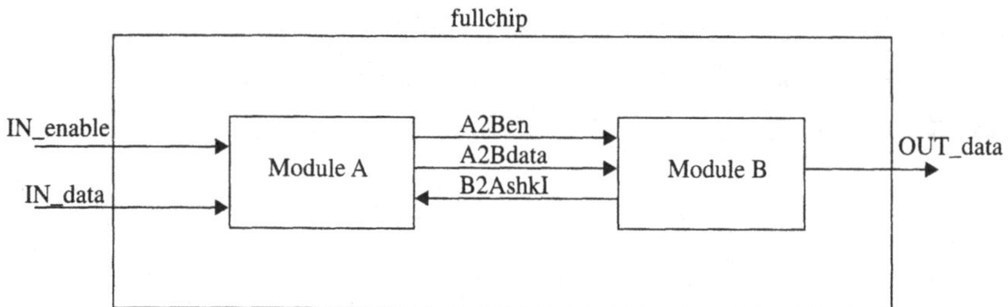

FIGURE 4.1. Diagram showing two submodules connected on a fullchip level.

Figure 4.1 is a diagram showing the fullchip level consisting of two modules, Module A and Module B. For this fullchip, let's assume the following rules for naming convention:

1. The first three characters of the signal name must be capitalized.
2. The first character must represent the name of the module on which the signal is an output from.
3. The third character must represent the name of the module on which the signal is an input to.
4. The second character must be the number 2.
5. The fourth character and beyond for a signal name is the signal name. It must be in lowercase letters.
6. Signals that propagate to the output at the fullchip level must have the first four characters "OUT_". The signal's name after the first four characters is in lowercase letters.
7. Signals that propagate from the input of the fullchip to a module must have the first three characters "IN_". The signal's name after the first three characters is in lowercase letters.
8. Any signal that is active low must end with the letter "I". The character must be in uppercase.

Based on these naming rules, the names of the signals "IN_enable," "IN_data," and "OUT_data" are input and output signals at the fullchip level. The names of signals that are interconnects between Module A and Module B are "A2Ben," "A2Bdata," and "B2AshkI."

The naming rules used here are just an example. A real design project may use naming rules that are similar to those shown here or they may be different.

Example 4.1 shows the Verilog code for Module A, Module B, and fullchip interconnect of both these modules.

Example 4.1 Verilog Example of Module A, Module B, and Fullchip Interconnect

```
module module_A (IN_enable, IN_data, B2AshkI, A2Ben,
A2Bdata);
input IN_enable, IN_data, B2AshkI;
output A2Ben, A2Bdata;
// your Verilog code for module_A
endmodule

module module_B (A2Ben, A2Bdata, B2AshkI, OUT_data);
input A2Ben, A2Bdata;
output B2AshkI, OUT_data;
// your Verilog code for module_B
endmodule

module fullchip (IN_enable, IN_data, OUT_data);
input IN_enable, IN_data;
output OUT_data;
wire A2Ben, A2Bdata, B2AshkI;
module_A module_A_instance (IN_enable, IN_data,
B2AshkI, A2Ben, A2Bdata);
module_B module_B_instance (A2Ben, A2Bdata, B2AshkI,
OUT_data);
endmodule
```

4.2 DESIGN PARTITIONING

It is good practice for the designer to partition a design into different modules. Each module should be partitioned with its own set of functionality or features. By having a good partitioning, the designer is able to break a complex design into smaller modules, thus giving more manageability to those modules. By following this method, the designer is able to localize the functionality of each module and write the HDL code for each module individually.

However, the designer needs to be careful when partitioning a design. Each module cannot be too small or too large. Partitioning modules that are too small will not be able to yield good synthesis optimization. Modules that are too large are difficult to be coded as well as synthesized to obtain optimal synthesis results. An acceptable module size that is manageable and allows good synthesis optimization, from a coding standpoint, would be around 5,000 to 15,000 gates.

Another point to keep in mind during design partitioning is the creation of additional interblock signaling. By partitioning into many blocks, a situation may occur whereby a need arises to create more signals for interfacing between these blocks. These additional signals may cause congestion in the layout phase as too many routing tracks are required. Therefore, it is important for the designer to fully understand the architecture and microarchitecture of a design before attempting to partition a design. Good partitioning brings advantages such as ease of manageability on each block. Bad partitioning brings disadvantages such as congestion on routing and increasing the die-size. Bad partitioning of a design also makes manageability of the design blocks a lot harder.

4.3 CLOCK

Most ASIC designs consist of at least one clock. Some may have more than one, some will only have one. Whether a design is a single clock design or multi-clock design, the designer needs to treat these clocks as global clocks. Global means that each clock is routed across all modules in the design, with the clock signal originating from a clock module.

A clock module that generates the global clock (or clocks) is nonsynthesizable. It is designed using conventional schematic capture. Analog blocks are integrated with the other logic blocks at the fullchip level.

Note: Analog blocks cannot be synthesized. In ASIC design flow, analog blocks are designed independently and integrated with logic blocks during fullchip integration. The reader must take note that only logic blocks can be coded into synthesizable HDL.

Referring to Figure 4.2, module A to module F are synthesizable logic modules. Each module is coded in HDL, verified using HDL testbench, and

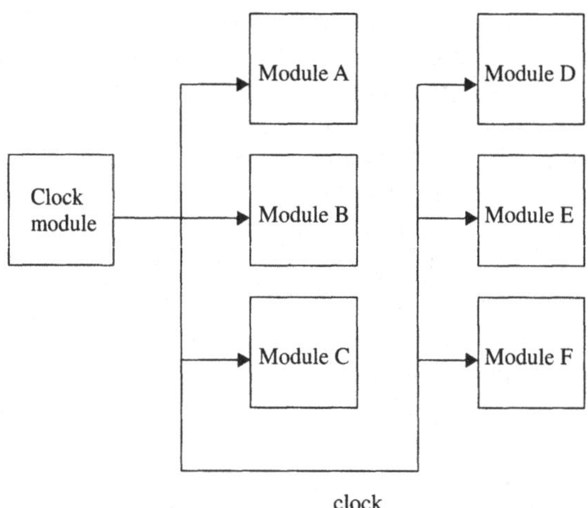

FIGURE 4.2. Diagram showing a fullchip level of global clock interconnect.

synthesized. During fullchip integration, the analog clock module is connected to the other logic modules.

When the designer codes logic modules A to F, he/she assumes the clock input is a global clock input that is able to meet the required clock skew. The global clock input is also assumed to have the clock period that is specified in the design specification. With these assumptions in mind, the designer cannot buffer the clock signal internally in the module. In other words, the clock signal must be considered as golden.

Treating the clock signal as a golden signal is a good practice when it comes to good coding style in HDL. By not buffering the clock signal, the designer is making the assumption that the clock signal is able to meet all the required specifications, which might not be the case. Clock skew is dependent on placement of cells (that has clock connected to it) and routing of the clock signals. Therefore, during coding and synthesis, clock should always be treated as golden, meaning that no buffering of any kind can be done on a clock during coding and synthesis. Any buffering on the clock signal to fix the clock skew should only be performed during clock tree synthesis (clock tree synthesis can be considered as part of APR in ASIC design methodology flow). Furthermore, tweaking of the clock signal to obtain the required clock period and clock duty cycle affects only the analog clock module, not the logic blocks of module A to F.

Figure 4.3 shows a diagram of an ideal design condition whereby the clock signal is directly connected to the clock's ports of the flip-flops used in the design without any logic gates or buffering on the clock signal. The designer should try to achieve this ideal condition in HDL coding whenever possible.

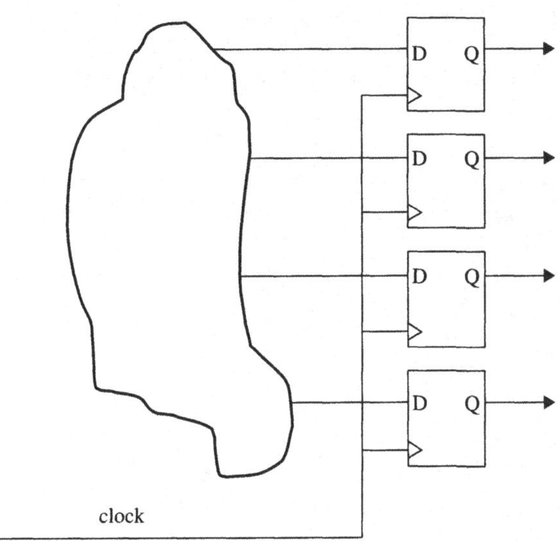

FIGURE 4.3. Diagram showing ideal connectivity of clock signal in a design.

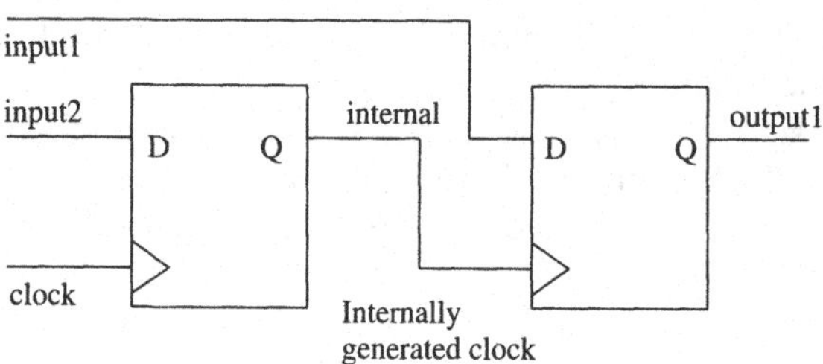

FIGURE 4.4. Diagram showing an output flip-flop driving another flip-flop.

The main advantage of having such a design is to allow the APR tool to perform its clock tree synthesis and insert clock buffers into the clock tree if needed. By doing so, the variable of clock skew can be ignored by the designer during HDL coding phase.

4.3.1 Internally Generated Clock

An internally generated clock should be used as little as possible. Ideally, synthesized designs should not have clocks that are internally generated.

Having synthesized designs that have flops or latches that are clocked internally complicates timing analysis. It is difficult to constrain the internal generated clock signal during synthesis.

Figure 4.4 is a diagram showing the output of a flip-flop being used to clock another flip-flop. Such a design can complicate the timing constraint process. Most synthesis and timing analysis tools have difficulties in identifying the type of internally generated clock design in Figure 4.4. Example 4.2 shows the Verilog code for the design in Figure 4.4.

Example 4.2 Verilog Code for the Design of Figure 4.4

```
module internal_clock (input1, input2, clock, output1);
input input1, input2, clock;
output output1;
reg internal,output1;

always @ (posedge clock)
begin
     internal <= input1;
end

always @ (posedge internal)
begin
     output1 <= input2;
end
endmodule
```

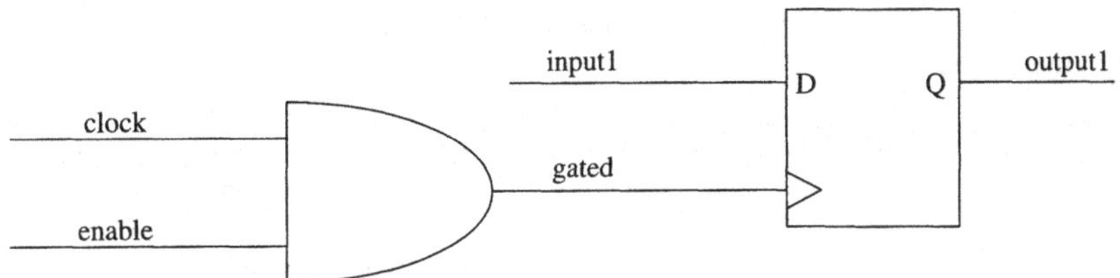

FIGURE 4.5. Diagram showing a gated clock driving a flip-flop.

4.3.2 Gated Clock

A design that has an enable signal to enable an internal clock, based on a global clock is called "gated clock." The term refers to the fact that the global clock is gated with a signal to generate the internal clock.

Gated clock designs are normally used when a designer wishes to switch off the clock signal under certain conditions. This could be for the purpose of power-saving features. Figure 4.5 shows an example of a design that has a flip-flop being clocked by a gated clock signal generated from an AND condition of "enable" and "clock."

Referring to the example in Figure 4.5, several ways can be used to code the design. The most common method is using boolean assignment and gate instantiation. Example 4.3 shows the Verilog code using boolean assignment. Example 4.4 shows the Verilog code using the gate instantiation method.

Example 4.3 Verilog Code for Gated Clock Design Using BOOLEAN Assignment

```
module gated_clock (input1, enable, clock, output1);
input input1, clock, enable;
output output1;
wire gated;
reg output1;

assign gated = clock & enable;
always @ (posedge gated)
begin
   output1 <= input1;
end

endmodule
```

Example 4.4 Verilog Code for Gated Clock Design Using Gate Instantiation

```
module gated_clock (input1, enable, clock, output1);
input input1, clock, enable;
output output1;
wire gated;
reg output1;

AND_gate AND_instance (.I1(clock), .I2(enable),
.O(gated));

always @ (posedge gated)
begin
     output1 <= input1;
end

endmodule
```

Verilog module **gated_clock** assumes a precompiled module **AND_gate.**

Note: Example 4.4 assumes of a precompiled ***AND*** *gate with inputs "**I1**" and "**I2**" and output "**O**". Another method to instantiate an* ***AND*** *gate is to use the built-in Verilog primitive "**and**" (refer to Section 3.2):* ***and AND_instance (O, I1, I2).***

Of the two methods, gate instantiation is the preferred method to handle gated clock. This is advisable, as instantiating the gate for the gated clock would allow the designer more control on the fanout of the signal ***gated***. For example, let's assume that the signal ***gated*** is to drive the clock of 32 flip-flops (Figure 4.6).

Referring to Figure 4.6, a fanout of 32 on signal ***gated*** is most likely to create a loading that is too heavy on the ***AND*** gate. As a result, the skew on the signal ***gated*** may be too large. Of course the designer can buffer up the signal ***gated*** during synthesis. However, buffering the signal ***gated*** is not recommended because it is a clock signal. Therefore, any buffering on the signal ***gated*** should be done only in APR (auto-place-route).

Therefore, a better approach would be the gate instantiation method. This method allows the designer to control the loading on the ***AND*** gate that drives the signal ***gated***. Using the same example of Figure 4.6, the designer can break the signal ***gated*** into several signals. And each signal would drive only a limited amount of flip-flops (refer to Fig. 4.7).

Referring to Figure 4.7, the signal ***clock*** and signal ***enable*** are used to create eight separate signals ***gated***, ranging from ***gated1*** to ***gated8***. Each ***gated*** signal only drives four flip-flops. To achieve this, the designer instantiates eight separate ***AND*** gates to create eight different ***gated*** signals. This method reduces the loading on each of the ***gated*** signal and allows the designer to achieve the required clock skew on signal ***gated***. Example 4.5 shows the Verilog code for

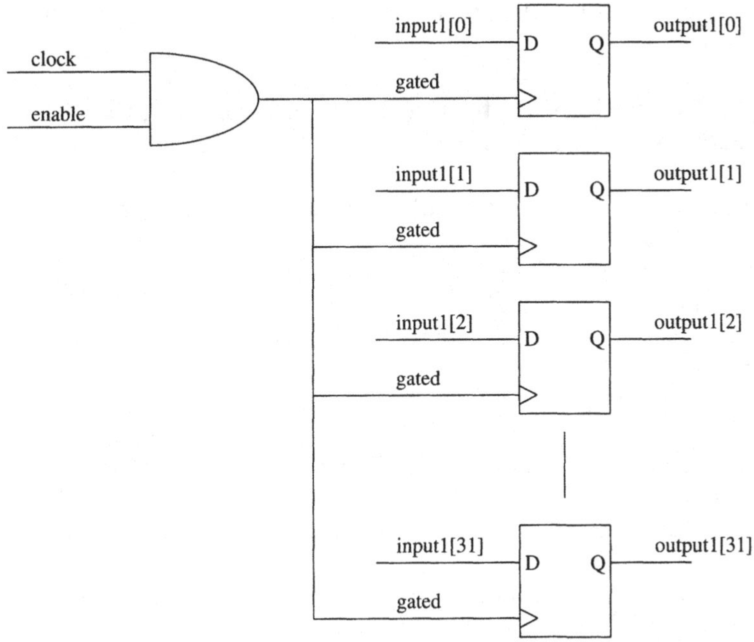

FIGURE 4.6. Diagram showing signal *gated* driving clock of 32 flip-flops.

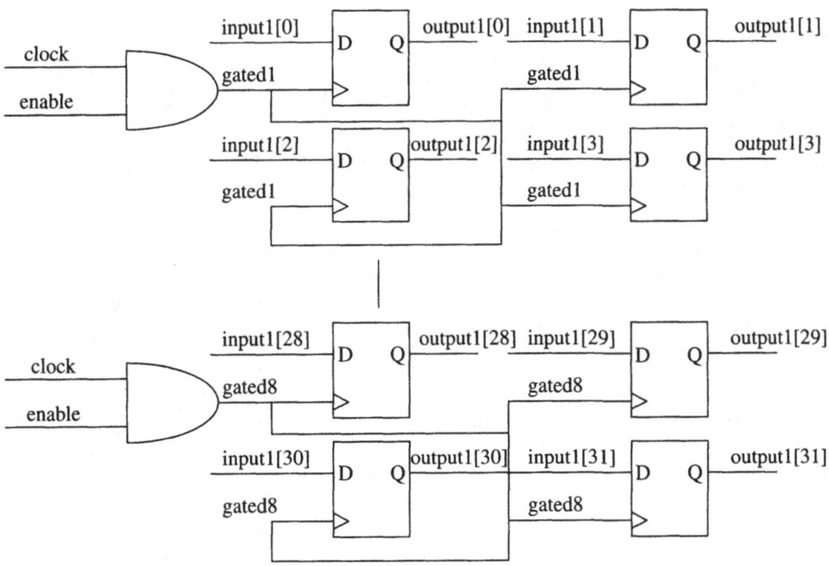

FIGURE 4.7. Diagram showing multiple *gated* signal to drive 32 flip-flops.

the gate instantiation method that allows a controlled clock skew on the signal ***gated*** (Fig. 4.7).

Example 4.5 Verilog Code for Gated Clock Design Using Gate Instantiation to Drive 32 Flip-flops

```
module gated_clock (input1, enable, clock, output1);
input [31:0] input1;
input clock, enable;
output [31:0] output1;
wire gated1, gated2, gated3, gated4, gated5, gated6,
gated7, gated8;
reg [31:0] output1;

AND_gate AND_instance1 (.I1(clock), .I2(enable),
.O(gated1));
AND_gate AND_instance2 (.I1(clock), .I2(enable),
.O(gated2));
AND_gate AND_instance3 (.I1(clock), .I2(enable),
.O(gated3));
AND_gate AND_instance4 (.I1(clock), .I2(enable),
.O(gated4));
AND_gate AND_instance5 (.I1(clock), .I2(enable),
.O(gated5));
AND_gate AND_instance6 (.I1(clock), .I2(enable),
.O(gated6));
AND_gate AND_instance7 (.I1(clock), .I2(enable),
.O(gated7));
AND_gate AND_instance8 (.I1(clock), .I2(enable),
.O(gated8));

always @ (posedge gated1)
begin
     output1[3:0] <= input1[3:0];
end

always @ (posedge gated2)
begin
     output1[7:4] <= input1[7:4];
end

always @ (posedge gated3)
begin
     output1[11:8] <= input1[11:8];
end
```

> Verilog module ***gated_clock*** assumes of a precompiled module ***AND_gate***. Another method is to use the built-in Verilog primitive "**and**".

```
always @ (posedge gated4)
begin
     output1[15:12] <= input1[15:12];
end

always @ (posedge gated5)
begin
     output1[19:16] <= input1[19:16];
end

always @ (posedge gated6)
begin
     output1[23:20] <= input1[23:20];
end

always @ (posedge gated7)
begin
     output1[27:24] <= input1[27:24];
end

always @ (posedge gated8)
begin
     output1[31:28] <= input1[31:28];
end

endmodule
```

4.4 RESET

Every design has some form of reset. It is a common requirement to allow a design to be "reset" to a certain known state during certain conditions.

There are two types of reset: asynchronous reset and synchronous reset. Both reset a design but their implications are very different.

4.4.1 Asynchronous Reset

Asynchronous reset is a reset that can occur at anytime. There is no reference of timing on this reset to any other signal. It can occur independent of any condition or other signal values. Figure 4.8 shows a simple design with a reset flip-flop. The output value of the flip-flop is a logical zero whenever the reset of the flip-flop is at a logical one. Example 4.6 is the Verilog code for a design example of asynchronous reset.

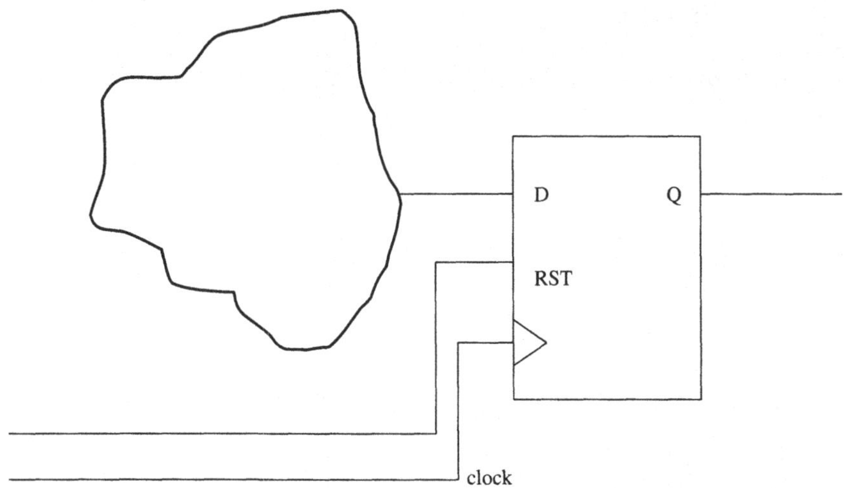

FIGURE 4.8. Diagram showing a design with an asynchronous reset flip-flop.

Example 4.6 Verilog Code for an Asynchronous Reset Design

```
module asynchronous_reset (input1, reset, clock,
output1);
input input1, reset, clock;
output output1;
reg output1;

always @ (posedge clock or posedge reset)
begin
     if (reset)
          output1 <= 1'b0;
     else
          output1 <= input1;
end

endmodule
```

Rising edge of ***reset***

4.4.2 Synchronous Reset

Synchronous reset is a reset that can only occur at the rising edge of clock for a positive clock-triggered flip-flop and falling edge of clock for a negative clock-triggered flip-flop. This means that synchronous reset is only recognized during rising edge or falling edge of clock. In other words, synchronous reset is referenced to the clock signal. It cannot occur independent of the clock. Figure 4.9 shows a simple synchronous reset design. The output value of the

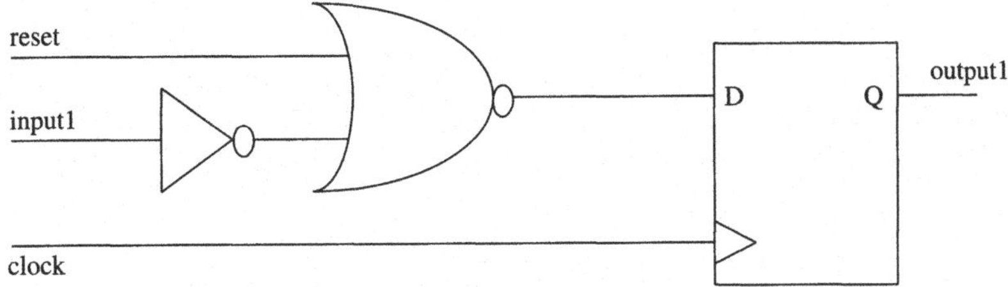

FIGURE 4.9. Diagram showing a design with a synchronous reset flip-flop.

flip-flop is updated during the rising edge of clock. The output value of the flip-flop is a logical zero, if during the rising edge of clock, reset is at logical one. The output value of the flip-flop is the logical value of the input data of the flip-flop, if reset is at logical zero during rising edge of clock. Example 4.7 shows the Verilog code for a design example of synchronous reset.

Example 4.7 Verilog Code for a Synchronous Reset Design

```
module synchronous_reset (input1, reset, clock,
output1);
input input1, reset, clock;
output output1;
reg output1;

always @ (posedge clock)
begin
     if (reset)
          output1 <= 1'b0;
     else
          output1 <= input1;
end

endmodule
```

Positive triggered flop with a synchronous reset

4.5 TIMING LOOP

Timing loops are loops in a design that have an output from combinational logic being looped back to be part of the input of the combinational logic.

For designs that are synthesized, it is important that they do not have timing loops. If a design has such loops, timing analysis is made complicated because the output is being looped back to the input. Figure 4.10 shows an example of a design having timing loop. Notice how the output of the inverter is being looped back as an input to the AND gate.

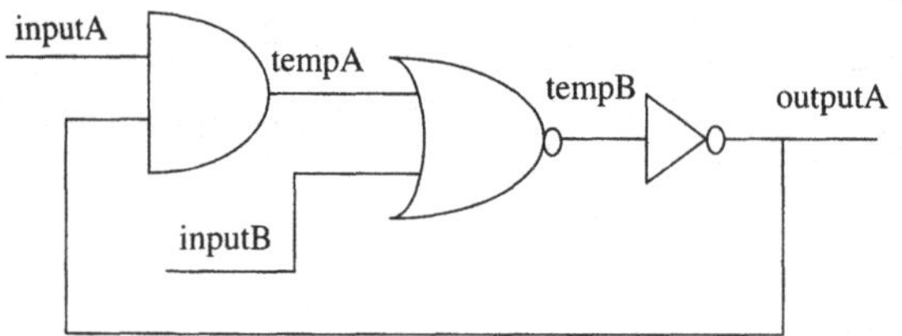

FIGURE 4.10. Diagram showing a design with timing loop.

When a design has timing loops, it is advisable that it be broken by a sequential element. This ensures that the timing loop, which may cause timing glitches, is broken into two timing paths: presequential and postsequential element path.

*Note: The Verilog code for the logic circuit in Figure 4.10 uses **outputA** to be looped back to generate **tempA**.*

```
module timingloop (inputA, inputB, outputA);
input inputA, inputB;
output outputA;
wire tempA, tempB;
assign tempA = inputA & outputA;
assign tempB = ~ (inputB | tempA);
assign outputA = ~tempB;
endmodule
```

It is not advisable for a designer to write Verilog code that uses timing loop. Logic circuits that have timing loops complicate timing analysis and have potential for causing timing glitches.

4.6 BLOCKING AND NONBLOCKING STATEMENTS

Blocking and nonblocking are two types of procedural assignments that are used in Verilog coding. Both of these types are used in sequential statements. Each of these blocking and nonblocking statements have different characteristics and behaviors.

Blocking statements are represented by the symbol "=". When a blocking statement is used, the statement is executed before the simulator moves forward to the next statement. In other words, a blocking statement is truly sequential.

Nonblocking statements are represented by the symbol "<=". When a nonblocking statement is used, that statement is scheduled and executed together with the other nonblocking assignments. What this means is that nonblocking allows several assignments to be scheduled and executed together, resulting in nonblocking statements that do not have dependence on the order in which the assignments occur (Examples 4.8 to 4.15 explain the difference between blocking and nonblocking statements in detail).

Do note that blocking and nonblocking statements refer only to Verilog code. VHDL code does not require concept of blocking and nonblocking.

Example 4.8 shows Verilog code for a simple design using nonblocking statements. The module "***non-blocking***" is basically a synchronous reset register-based design.

Example 4.8 Verilog Code Showing Use of Nonblocking Statement

```
module nonblocking (clock, input1, reset, output1,
output2, output3);
input reset, clock;
input [3:0] input1;
output [3:0] output1, output2, output3;

always @ (posedge clock)
begin
     if (reset)
          begin
               output1 <= 4'b0000;
               output2 <= 4'b0000;
               output3 <= 4'b0000;
          end
     else
          begin
               output1 <= input1;
               output2 <= output1;
               output3 <= output2;
          end
end
endmodule
```

Using the Verilog code of Example 4.8, Example 4.9 and Example 4.10 shows two other Verilog codes using nonblocking statements. Each Verilog code of Example 4.8, 4.9, and 4.10 uses a different order to assign values to ***output1***, ***output2***, and ***output3***.

Example 4.9 Verilog Code for Example 4.8 with the Output Assignment Rearranged

```
module nonblocking (clock, input1, reset, output1,
output2, output3);
```

```
input reset, clock;
input [3:0] input1;
output [3:0] output1, output2, output3;

always @ (posedge clock)
begin
    if (reset)
        begin
            output1 <= 4'b0000;
            output2 <= 4'b0000;
            output3 <= 4'b0000;
        end
    else
        begin
            output1 <= input1;
            output3 <= output2;
            output2 <= output1;
        end
end
endmodule
```

Assignment of **output1, output2** and **output3** rearranged.

Example 4.10 Verilog Code for Example 4.9 with the Output Assignment Rearranged

```
module nonblocking (clock, input1, reset, output1,
output2, output3);
input reset, clock;
input [3:0] input1;
output [3:0] output1, output2, output3;

always @ (posedge clock)
begin
    if (reset)
        begin
            output1 <= 4'b0000;
            output2 <= 4'b0000;
            output3 <= 4'b0000;
        end
    else
        begin
            output2 <= output1;
            output3 <= output2;
            output1 <= input1;
        end
end
endmodule
```

Assignment of **output1, output2** and **output3** rearranged.

Notice how the Verilog code of Examples 4.8, 4.9, and 4.10 are basically the same except for the arrangement of sequence of assignment for ***output1***, ***output2***, and ***output3***. A simple test bench is written to simulate all three of these examples. Example 4.11 shows the Verilog code for the test bench.

Example 4.11 Verilog Code for Test Bench for Simulation of Examples 4.8, 4.9, and 4.10

```
module nonblocking_tb ();

reg [3:0] input1;
reg clock, reset;
wire [3:0] output1, output2, output3;

initial
begin
     clock = 0;
     input1 = 0;
     forever #50 clock = ~clock;
end

initial
begin
     #10;
     reset = 0;
     #10;
     reset = 1;
     #10;
     reset = 0;
     #10;
     input1 = 1;
     #50;
     input1 = 2;
     #200;
     $finish;
end

nonblocking nonblocking_instance (clock, input1, reset,
output1, output2, output3);

endmodule
```

Using the test bench shown in Example 4.11, the Verilog code for Examples 4.8, 4.9, and 4.10 is simulated. The simulation results are shown in Figures 4.11, 4.12, and 4.13.

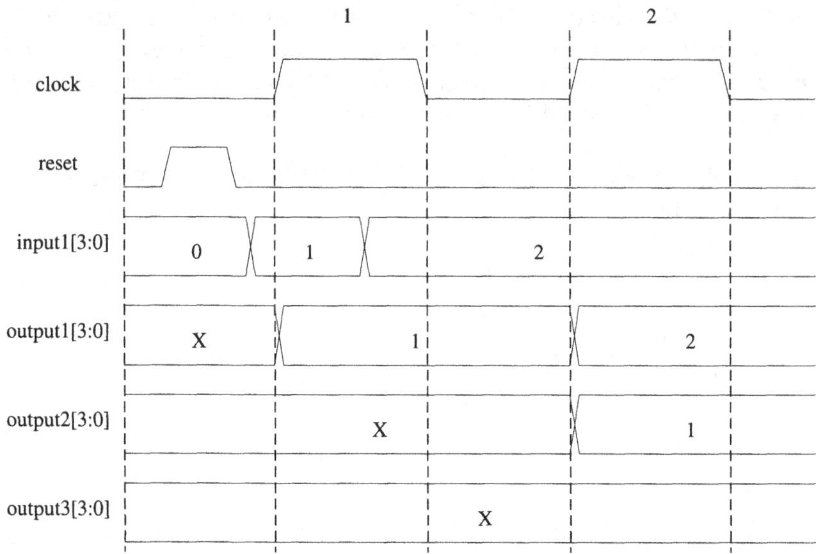

FIGURE 4.11. Diagram showing simulation results of Verilog code in Example 4.8.

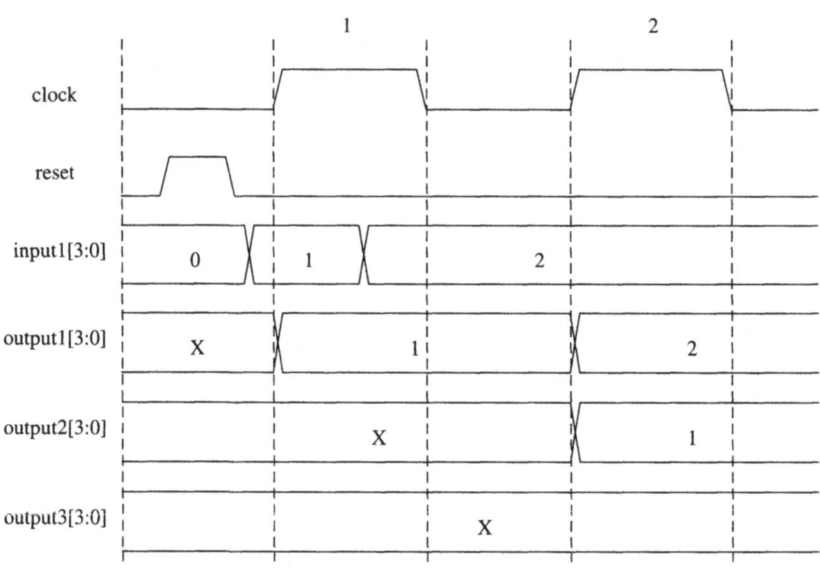

FIGURE 4.12. Diagram showing simulation results of Verilog code in Example 4.9.

Notice the simulation results shown in Figures 4.11, 4.12, and 4.13 are the same. Although the sequence of statement assignments of ***output1***, ***output2***, and ***output3*** are different for the Verilog code of Examples 4.8, 4.9, and 4.10, the simulation results are the same. Changing the arrangement of the sequence

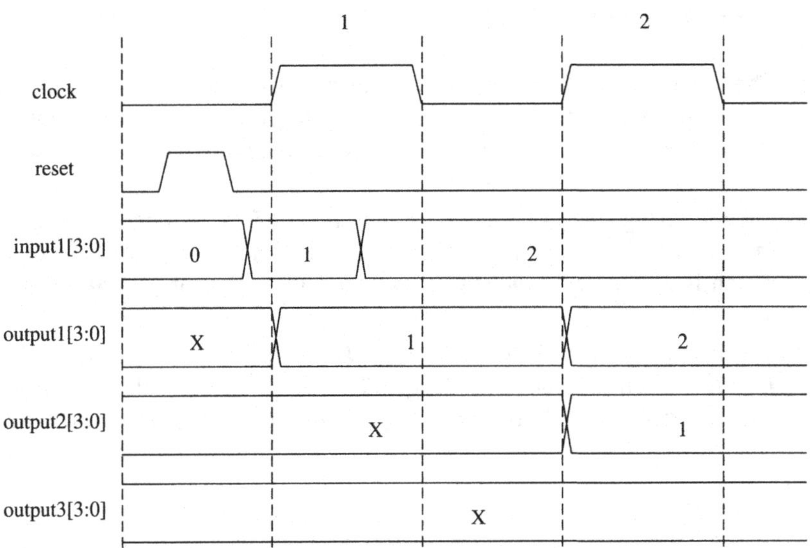

FIGURE 4.13. Diagram showing simulation results of Verilog code in Example 4.10.

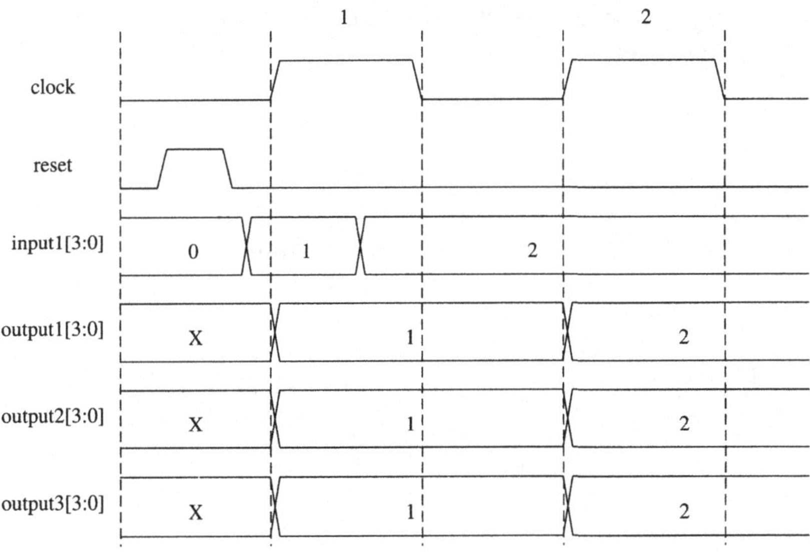

FIGURE 4.14. Diagram showing simulation results of Verilog code in Example 4.12.

of statement assignments does not affect simulation because the three examples uses nonblocking statements. This would mean that the assignment of "***output1 ⇐ input1***," "***output2 ⇐ output1***," and "***output3 ⇐ output2***" are executed together. Therefore, when using nonblocking statements, order dependence does not affect simulation results.

Referring to Figures 4.11, 4.12, and 4.13,

1. The Verilog codes in Examples 4.8, 4.9, and 4.10 use a synchronous reset. Therefore, when reset is at logical “1”, ***output1***, ***output2***, and ***output3*** are not reset to a value of “0” because a positive edge of ***clock*** did not occur.
2. On the rising edge of the first ***clock***, ***output1*** is assigned the value of ***input1***, which is decimal 1. ***Output2*** is assigned the value of ***output1***, which is *X*. Similarly ***output3*** is assigned the value of ***output2***, which is also *X*.
3. On the rising edge of the second ***clock***, ***output1*** is assigned the value of ***input1***, which is decimal 2. ***Output2*** is assigned the value of ***output1***, which is decimal 1. And ***output3*** is assigned the value of ***output2***, which is *X*.

Examples 4.12, 4.13, and 4.14 are the same pieces of Verilog code as Examples 4.8, 4.9, and 4.10, but blocking statements are used instead of nonblocking statements.

Example 4.12 Verilog Code Showing Example 4.8 Using a Blocking Statement

```
module blocking (clock, input1, reset, output1,
output2, output3);
input reset, clock;
input [3:0] input1;
output [3:0] output1, output2, output3;

always @ (posedge clock)
begin
     if (reset)
          begin
               output1 = 4'b0000;
               output2 = 4'b0000;
               output3 = 4'b0000;
          end
     else
          begin
               output1 = input1;
               output2 = output1;
               output3 = output2;
          end
end
endmodule
```

Example 4.13 Verilog Code Showing Example 4.9 Using a Blocking Statement

```
module blocking (clock, input1, reset, output1,
output2, output3);
input reset, clock;
input [3:0] input1;
output [3:0] output1, output2, output3;

always @ (posedge clock)
begin
    if (reset)
        begin
            output1 = 4'b0000;
            output2 = 4'b0000;
            output3 = 4'b0000;
        end
    else
        begin
            output1 = input1;
            output3 = output2;
            output2 = output1;
        end
end
endmodule
```

Assignment of ***output1***, ***output2*** and ***output3*** rearranged.

Example 4.14 Verilog Code Showing Example 4.10 Using Blocking Statement

```
module blocking (clock, input1, reset, output1,
output2, output3);
input reset, clock;
input [3:0] input1;
output [3:0] output1, output2, output3;

always @ (posedge clock)
begin
    if (reset)
        begin
            output1 = 4'b0000;
            output2 = 4'b0000;
            output3 = 4'b0000;
        end
    else
        begin
```

```
                output2 = output1;
                output3 = output2;
                output1 = input1;
            end
end
endmodule
```

Assignment of ***output1, output2*** and ***output3*** rearranged.

Notice how the Verilog code of Examples 4.12, 4.13, and 4.14 are basically the same except for the arrangement of sequence of assignment for ***output1***, ***output2***, and ***output3***. A testbench is written to simulate all three of these examples. Example 4.15 shows the Verilog code for the testbench. Also take note that the stimulus used in the testbench of Example 4.15 is the same set of stimulus used in the testbench of Example 4.11.

Example 4.15 Verilog Code for Testbench for Simulation of Examples 4.12, 4.13, and 4.14

```
module blocking_tb ();

reg [3:0] input1;
reg clock, reset;
wire [3:0] output1, output2, output3;

initial
begin
    clock = 0;
    input1 = 0;
    forever #50 clock = ~clock;
end

initial
begin
    #10;
    reset = 0;
    #10;
    reset = 1;
    #10;
    reset = 0;
    #10;
    input1 = 1;
    #50;
    input1 = 2;
    #200;
    $finish;
end
```

```
blocking blocking_instance (clock, input1, reset,
output1, output2, output3);

endmodule
```

Using the test bench shown in Example 4.15, the Verilog code for Examples 4.12, 4.13, and 4.14 are simulated. The simulation results are shown in Figures 4.14, 4.15, and 4.16.

Referring to the simulation waveform in Figure 4.14:

1. The Verilog code in Example 4.12 uses a synchronous reset. Therefore, when the reset is at logical "1", ***output1***, ***output2***, and ***output3*** are not reset to value of "0" because a positive edge of ***clock*** did not occur.
2. On the rising edge of the first ***clock***, ***output1*** is assigned the value of ***input1***, which is decimal 1. ***Output2*** is assigned the value of ***output1***. Because this is a blocking statement, the assignment of "***output2 = output1***" will only occur after the assignment of "***output1 = input1***" has completed. Thus, ***output2*** has a value of decimal 1. Similarly, the assignment of "***output3 = output2***" only occurs after the assignment of "***output2 = output1***." Because ***output2*** has a value of decimal 1, ***output3*** is also assigned a value of decimal 1.
3. On the rising edge of the second ***clock***, ***output1*** is assigned the value of ***input1***, which is a decimal 2. ***Output2*** is assigned the value of ***output1***.

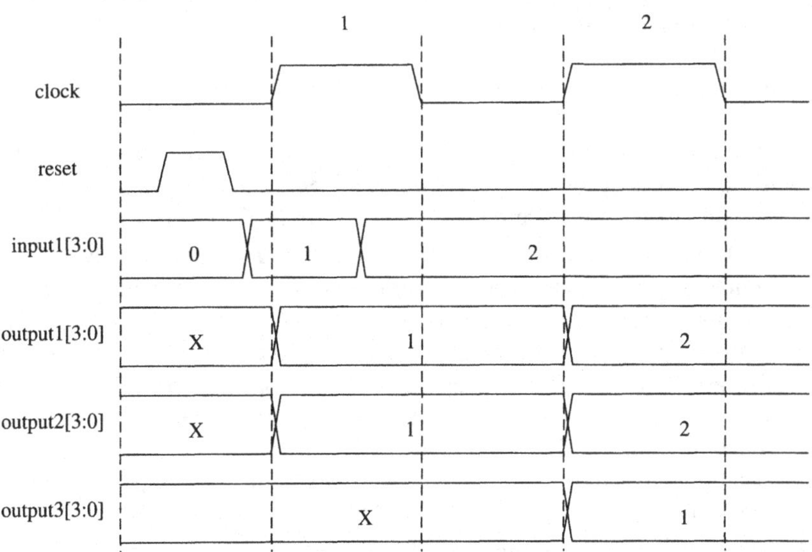

FIGURE 4.15. Diagram showing simulation results of Verilog code in Example 4.13.

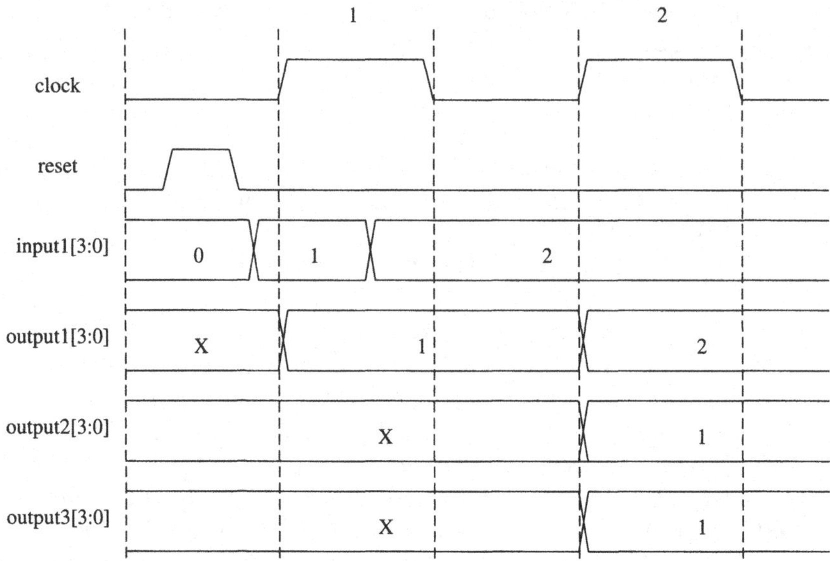

FIGURE 4.16. Diagram showing simulation results of Verilog code in Example 4.14.

Because this is a blocking statement, the assignment of "***output2* = *output1***" will only occur after the assignment of "***output1* = *input1***" has completed. Thus, ***output2*** has a value of decimal 2. Similarly, the assignment of "***output3* = *output2***" only occurs after the assignment of "***output2* = *output1***." Because ***output2*** has a value of decimal 2, ***output3*** is also assigned a value of decimal 2.

Referring to the simulation waveform in Figure 4.15:

1. The Verilog code in Example 4.13 uses a synchronous reset. Therefore, when reset is at logical "1," ***output1***, ***output2***, and ***output3*** are not reset to value of "0" because a positive edge of ***clock*** did not occur.
2. On the rising edge of the first ***clock***, ***output1*** is assigned the value of ***input1***, which is a decimal 1. ***Output3*** is assigned the value of ***output2***. Because this is a blocking statement, at the moment when the assignment of "***output3* = *output2***" occurs, ***output2*** is at X. Thus, ***output3*** is also assigned X. For the assignment of "***output2* = *output1***," this blocking statement occurs after the assignments of "***output1* = *input1***" and "***output3* = *output2***" have occurred. Because ***output1*** has already been assigned the value of decimal 1, ***output2*** is also assigned the value of decimal 1.
3. On the rising edge of the second ***clock***, ***output1*** is assigned the value of ***input1***, which is a decimal 2. ***Output3*** is assigned the value of ***output2***,

which is a decimal 1. For the assignment of "***output2 = output1***," which is a blocking statement, this statement assignment only occurs after "***output1 = input1***" and "***output3 = output2***" have occurred. Therefore, ***output2*** is assigned the value of decimal 1.

Referring to the simulation waveform in Figure 4.16:

1. The Verilog code in Example 4.14 uses a synchronous reset. Therefore, when reset is at logical "1," ***output1***, ***output2***, and ***output3*** are not reset to value of "0" because a positive edge of ***clock*** did not occur.
2. On the rising edge of the first ***clock***, ***output2*** is assigned the value of ***output1***, which is X. ***Output3*** is assigned the value of ***output2***. At the moment when the assignment of "***output3 = output2***" occurs, ***output2*** is at X. Therefore, ***output3*** is also assigned X. For the assignment of "***output1 = input1***," because ***input1*** has a value of decimal 1, ***output1*** is also assigned the value of decimal 1.
3. On the rising edge of the second ***clock***, ***output2*** is assigned value of ***output1***, which is a decimal 1. ***Output3*** is assigned the value of ***output2***, which is a decimal 1. For the assignment of "***output1 = input1***," because ***input1*** has a value of decimal 2, ***output1*** is assigned the value of decimal 2.

Referring to simulation waveforms of Examples 4.12, 4.13, and 4.14 (Figs. 4.14, 4.15, and 4.16), it is obvious that use of a blocking statement within an "***always @ (posedge***" block gives different simulation results when the order of statement assignment is changed. In other words, use of a blocking statement is order dependent.

Referring to simulation waveforms of Examples 4.8, 4.9, and 4.10 (Figs. 4.11, 4.12, and 4.13), use of a nonblocking statement within an "***always @ (posedge***" block gives the same simulation results when the order of statement assignment is changed. In other words, use of a nonblocking statement is not order dependent.

From the exercise of Examples 4.8 to 4.15, it is concluded that nonblocking statements must be used when coding for registers. When coding for combinational logic, blocking statements are used. This ensures that when a designer is writing code to synthesize registers, the order in which the nonblocking statements are written will not affect simulation.

Note: When writing Verilog code that involves more than one register assignment, always use nonblocking statements. This will ensure that the order in which the register assignments is made does not affect the simulation results.

4.7 SENSITIVITY LIST

Verilog uses a sensitivity list to determine if a block of sequential statements needs to be evaluated by the simulator during certain simulation cycles. For Verilog, a sensitivity list is required for the ***always*** statement.

Example 4.16 shows a Verilog code for a design module that has an "***always***" block. This block is to be evaluated by the simulator whenever there is a change in the signals corresponding to the sensitivity list of the "***always***" block.

Example 4.16 Verilog Example Showing Sensitivity List for "always" Block

```
module senselist (X, Y, Z, AB);
input X, Y, Z;
output AB;

always @ (X or Y or Z)———►  Sensitivity list which
begin                       consists of X, Y and Z.
    // design source code
end
endmodule
```

Referring to Example 4.16, the sensitivity list consists of three signals, ***X***, ***Y***, and ***Z***. The block of sequential statements within the "***always***" block is evaluated by the simulator whenever there is a change of values in either signal ***X***, ***Y***, or ***Z***.

An incomplete signal list in the sensitivity list for an "***always***" block may cause simulation results to be inaccurate. It may also cause a mismatch between the synthesis results of the Verilog code and the simulation results. It is therefore important to always keep note that signals evaluated in an "***always***" block needs to be included in the sensitivity list.

Table 4.1 is an example of the differences in simulation that may occur due to an incomplete senstivity list.

Notice how the simulation results for the modules differ. The result of ***outputA*** for the module that has incomplete sensitivity list has a logic value of "1" for all combinations of inputs ***inputA***, ***inputB***, and ***inputC***. The module that has complete sensitivity list has the results of ***outputA*** at logic "0" when inputs ***inputA***, ***inputB***, and ***inputC*** are at a combination of "111."

Both of the modules, although having different simulation results, when synthesized will generate a NAND gate. In this case, it is clear that that synthesized logic will never match the simulation result of the module with incomplete sensitivity list. It is therefore very important for a designer to always use a complete sensitivity list when using an ***always*** statement in Verilog.

TABLE 4.1. Differences in simulation resulting from an incomplete sensitivity list

Complete Sensitivity List	Incomplete Sensitivity List

module sense (inputA, inputB,
inputC, outputA);
 	
input inputA, inputB, inputC;
output outputA;
 	
reg outputA;
 	
always @ (inputA or inputB or inputC) begin
  if (inputA & inputB & inputC)
    outputA = 0;
  else
    outputA = 1;
end
 	
endmodule
	module sense (inputA, inputB,
	inputC, outputA);
	
	input inputA, inputB, inputC;
	output outputA;
	
	reg outputA;
	
	always @ (inputA or inputB) begin
	  if (inputA & inputB & inputC)
	    outputA = 0;
	  else
	    outputA = 1;
	end
	
	endmodule

Testbench to Simulate Both the Verilog Code

```
module sense_tb();

reg reg_inputA, reg_inputB, reg_inputC;
wire wire_outputA;

integer i;

initial
begin
  for (i = 0; i < 8; i = i + 1)
  begin
    {reg_inputA, reg_inputB, reg_inputC} = i;
    #100;
  end
end

sense sense_inst (.inputA(reg_inputA), .inputB(reg_inputB),
.inputC(reg_inputC), .outputA(wire_outputA));

initial
begin
  $monitor ("reg_inputA %b reg_inputB %b reg_inputC %b wire_outputA %b\n",
reg_inputA, reg_inputB, reg_inputC, wire_outputA);
end

endmodule
```

TABLE 4.1. (Continued)

Simulation Results (Complete Sensitivity List)				Simulation Results (Incomplete Sensitivity List)			
inputA	inputB	inputC	outputA	inputA	inputB	inputC	outputA
0	0	0	1	0	0	0	1
0	0	1	1	0	0	1	1
0	1	0	1	0	1	0	1
0	1	1	1	0	1	1	1
1	0	0	1	1	0	0	1
1	0	1	1	1	0	1	1
1	1	0	1	1	1	0	1
1	1	1	0	1	1	1	1
Synthesized Logic (Complete Sensitivity List) 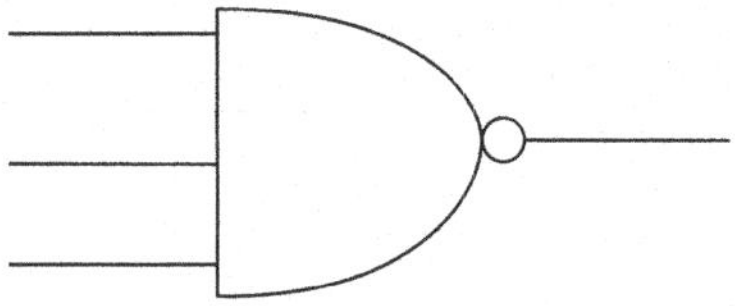				Synthesized Logic (Incomplete Sensitivity List) 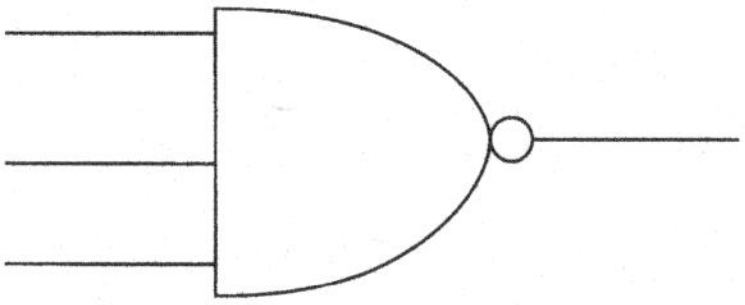			

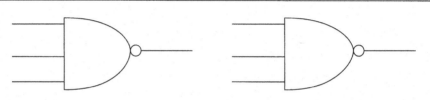

4.8 VERILOG OPERATORS

Verilog allows the use of a large number of operators. Operators form the very basic components when coding for a design. It allows the designer to use these operators to achieve different functionalities and operations.

All Verilog operators are synthesizable. These operators can be grouped into different types, with each type having its own set of functionality.

4.8.1 Conditional Operators

Conditional operators are commonly used to model combinational logic designs that behave as a switching device. A conditional operator consists of three operands: (a) the input expression; (b) the select control signal that selects which input expression is to be passed through to the output; (c) the output expression.

The syntax for a conditional operator is as follows:

```
assign output_signal = control_signal ? input1 :
input2;
```

whereby ***output_signal*** is the output of the conditional statement, ***control_signal*** is the signal that chooses whether ***input1*** or ***input2*** is passed

TABLE 4.2. Truth table showing functionality for module "conditional"

InputA	InputB	ControlC	OutputA
0	0	0	0
0	0	1	0
0	1	0	1
0	1	1	0
1	0	0	0
1	0	1	1
1	1	0	1
1	1	1	1

to **output_signal** (if ***control_signal*** is true, ***input1*** is passed to ***output_signal***, otherwise, ***input2***).

Table 4.2 is a truth table that shows the functionality of a module called "***conditional***," which can be modeled in Verilog using the conditional operator.

Example 4.17 shows the Verilog code for module "***conditional***," which has the functionality of Table 4.2.

Example 4.17 Verilog Code for Module "conditional"

```
module conditional (inputA, inputB, controlC, outputA);

input inputA, inputB, controlC;
output outputA;

wire outputA;

assign outputA = controlC ? inputA : inputB;

endmodule
```

When Example 4.17 is synthesized, Figure 4.17 is obtained. The logic synthesized from Example 4.17 is a multiplexer. Therefore, when coding for synthesis, a good method to code for multiplexers is to use conditional operators.

4.8.2 Bus Concatenation Operator

Multiple signals can be concatenated to form a bus. This can be achieved by using the bus concatenation operator. The syntax on using this operator is

```
assign signal_bus = {signal1, signal2, signal3};
```

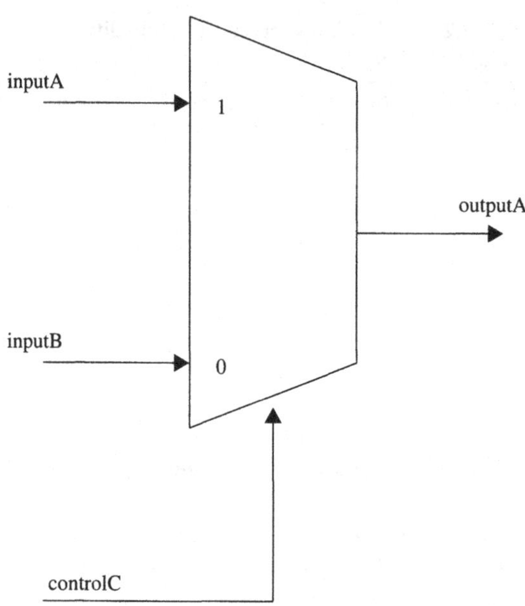

FIGURE 4.17. Diagram showing synthesized logic for module "*conditional.*"

whereby ***signal_bus*** is the name of the three-bit concatenated bus and ***signal1***, ***signal2***, and ***signal3*** are the signals concatenated together.

Example 4.18 Verilog Example Showing a Three-Bit and Four-Bit Bus Concatenation

```
module concatenate (inputA, inputB, inputC, inputD,
outputA, outputB);

input inputA, inputB, inputC, inputD;
output [2:0] outputA;
output [3:0] outputB;

wire [2:0] outputA;
wire [3:0] outputB;

assign outputA = {inputA, inputB,
inputC};
assign outputB = {inputA, inputB,
inputC, inputD};

endmodule
```

three-bit bus concatenation

four-bit bus concatenation

Example 4.18 shows a Verilog code that concatenates three signals, ***inputA***, ***inputB***, and ***inputC***, into a three-bit bus ***outputA*** and the concatenation of four signals, ***inputA***, ***inputB***, ***inputC***, and ***inputD*** into a four-bit bus ***outputB***.

4.8.3 Shift Operator

Shift operations can be performed in Verilog by using the shift left operator for shifting a bus to the left or a shift right operator for shifting a bus to the right.

Example 4.19 shows a Verilog code that uses the shift left operator to shift the three-bit bus signal ***tempA*** by one bit to the left.

Example 4.19 Verilog Code Using the Shift Left Operator

```
module shift_left (inputA, inputB, outputA);

input [2:0] inputA, inputB;    --> Declaration of inputA
output [2:0] outputA;              and inputB as three-
                                   bit input ports.

wire [2:0] outputA;

wire [2:0] tempA;

assign tempA = inputA & inputB;      Shift left by
                                     one bit
assign outputA = tempA << 1;

endmodule
```

When the Verilog code of Example 4.19 is synthesized, the logic obtained is illustrated in Figure 4.18.

Example 4.20 shows the Verilog code for a test bench that can be used to simulate the Verilog code of module "***shift_left***" to verify that the logic obtained is as shown in Figure 4.18.

Example 4.20 Verilog Code for Test Bench to Simulate Module "*shift_left*"

```
module shift_left_tb();

reg [2:0] reg_inputA, reg_inputB;
wire [2:0] wire_outputA;

integer i,j;
```

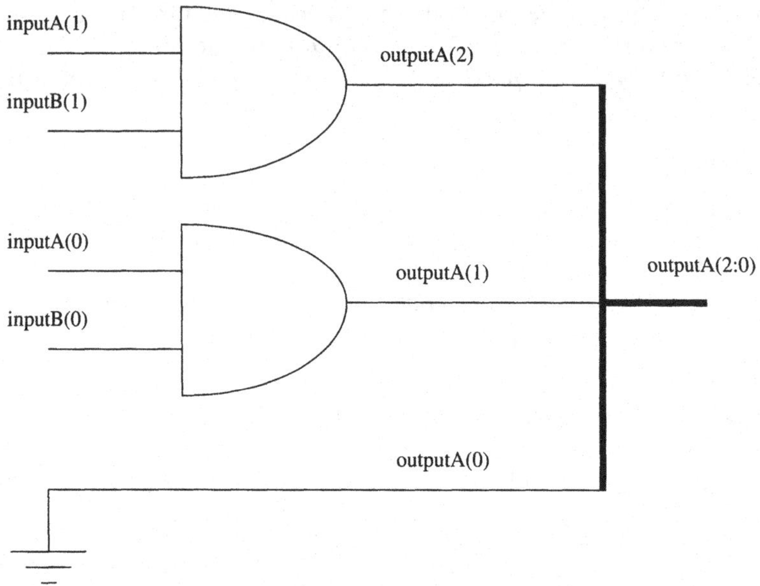

FIGURE 4.18. Diagram showing synthesized logic for module "***shift_left***."

```
initial
begin
   for (i=0; i<8; i=i+1)
      begin
         // to force input stimulus for inputA
         reg_inputA = i;
         for (j=0; j<8; j=j+1)
            begin
               // to force input stimulus for inputB
               reg_inputB = j;
               #10;
            end
      end
end

shift_left shift_left_inst (.inputA(reg_inputA),
.inputB(reg_inputB), .outputA(wire_outputA));

initial
begin
    $monitor ("inputA %b%b%b inputB %b%b%b tempA
%b%b%b outputA %b%b%b",reg_inputA[2], reg_inputA[1],
reg_inputA[0], reg_inputB[2], reg_inputB[1],
```

A for loop operation that allows a loop of i = 0 to i = 7

Instantiation of module ***shift_left***

```
reg_inputB[0], shift_left_inst.tempA[2],
shift_left_inst.tempA[1], shift_left_inst.tempA[0],
wire_outputA[2], wire_outputA[1], wire_outputA[0]);
end

endmodule
```

$monitor is a Verilog system task that allows the monitoring of simulation results during simulation. The task would display the values of the specified signals during simulation whenever the signal values change.

Example 4.21 shows the simulation results of the test bench module "***shift_left_tb***".

Example 4.21 Simulation Results of Test Bench Module "*shift_left_tb*"

```
inputA 000 inputB 000 tempA 000 outputA 000
inputA 000 inputB 001 tempA 000 outputA 000
inputA 000 inputB 010 tempA 000 outputA 000
inputA 000 inputB 011 tempA 000 outputA 000
inputA 000 inputB 100 tempA 000 outputA 000
inputA 000 inputB 101 tempA 000 outputA 000
inputA 000 inputB 110 tempA 000 outputA 000
inputA 000 inputB 111 tempA 000 outputA 000
inputA 001 inputB 000 tempA 000 outputA 000
inputA 001 inputB 001 tempA 001 outputA 010
inputA 001 inputB 010 tempA 000 outputA 000
inputA 001 inputB 011 tempA 001 outputA 010
inputA 001 inputB 100 tempA 000 outputA 000
inputA 001 inputB 101 tempA 001 outputA 010
inputA 001 inputB 110 tempA 000 outputA 000
inputA 001 inputB 111 tempA 001 outputA 010
inputA 010 inputB 000 tempA 000 outputA 000
inputA 010 inputB 001 tempA 000 outputA 000
inputA 010 inputB 010 tempA 010 outputA 100
inputA 010 inputB 011 tempA 010 outputA 100
inputA 010 inputB 100 tempA 000 outputA 000
inputA 010 inputB 101 tempA 000 outputA 000
inputA 010 inputB 110 tempA 010 outputA 100
inputA 010 inputB 111 tempA 010 outputA 100
```

```
inputA 011 inputB 000 tempA 000 outputA 000
inputA 011 inputB 001 tempA 001 outputA 010
inputA 011 inputB 010 tempA 010 outputA 100
inputA 011 inputB 011 tempA 011 outputA 110
inputA 011 inputB 100 tempA 000 outputA 000
inputA 011 inputB 101 tempA 001 outputA 010
inputA 011 inputB 110 tempA 010 outputA 100
inputA 011 inputB 111 tempA 011 outputA 110
inputA 100 inputB 000 tempA 000 outputA 000
inputA 100 inputB 001 tempA 000 outputA 000
inputA 100 inputB 010 tempA 000 outputA 000
inputA 100 inputB 011 tempA 000 outputA 000
inputA 100 inputB 100 tempA 100 outputA 000
inputA 100 inputB 101 tempA 100 outputA 000
inputA 100 inputB 110 tempA 100 outputA 000
inputA 100 inputB 111 tempA 100 outputA 000
inputA 101 inputB 000 tempA 000 outputA 000
inputA 101 inputB 001 tempA 001 outputA 010
inputA 101 inputB 010 tempA 000 outputA 000
inputA 101 inputB 011 tempA 001 outputA 010
inputA 101 inputB 100 tempA 100 outputA 000
inputA 101 inputB 101 tempA 101 outputA 010
inputA 101 inputB 110 tempA 100 outputA 000
inputA 101 inputB 111 tempA 101 outputA 010
inputA 110 inputB 000 tempA 000 outputA 000
inputA 110 inputB 001 tempA 000 outputA 000
inputA 110 inputB 010 tempA 010 outputA 100
inputA 110 inputB 011 tempA 010 outputA 100
inputA 110 inputB 100 tempA 100 outputA 000
inputA 110 inputB 101 tempA 100 outputA 000
inputA 110 inputB 110 tempA 110 outputA 100
inputA 110 inputB 111 tempA 110 outputA 100
inputA 111 inputB 000 tempA 000 outputA 000
inputA 111 inputB 001 tempA 001 outputA 010
inputA 111 inputB 010 tempA 010 outputA 100
inputA 111 inputB 011 tempA 011 outputA 110
inputA 111 inputB 100 tempA 100 outputA 000
inputA 111 inputB 101 tempA 101 outputA 010
inputA 111 inputB 110 tempA 110 outputA 100
inputA 111 inputB 111 tempA 111 outputA 110
```

Note: Notice from the simulation results that the (LSB) is always a zero? This occurs because, when shifting left, the LSB is always tagged with logic zero. This causes the synthesized logic for module ***shift_left*** *to have the* ***outputA(0)*** *grounded.*

Example 4.22 shows a Verilog code that uses the shift right operator to shift the three-bit bus signal ***tempA*** by one bit to the right.

Example 4.22 Verilog Code Using the Shift Right Operator

```
module shift_right (inputA, inputB, outputA);

input [2:0] inputA, inputB;
output [2:0] outputA;

wire [2:0] outputA;

wire [2:0] tempA;

assign tempA = inputA & inputB;

assign outputA = tempA >> 1;     Shift right by one bit

endmodule
```

When the Verilog code of Example 4.22 is synthesized, the logic obtained is illustrated in Figure 4.19.

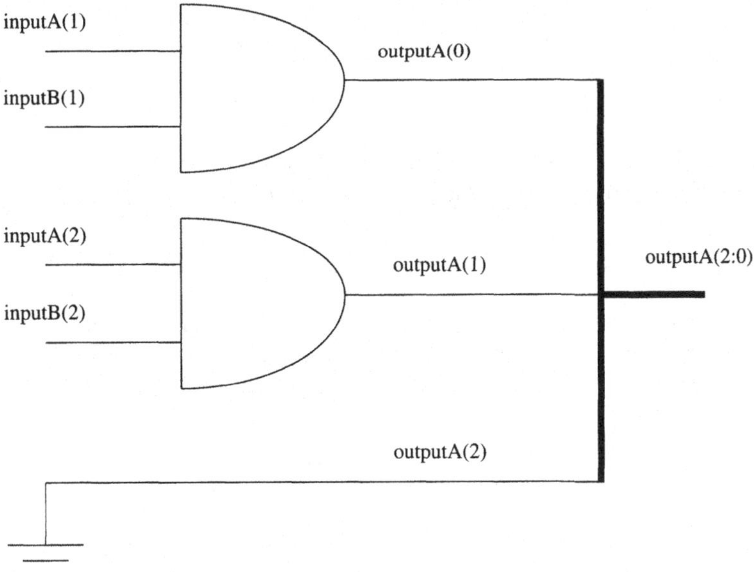

FIGURE 4.19. Diagram showing synthesized logic for module "***shift_right***."

Example 4.23 shows the Verilog code for a test bench that can be used to simulate the verilog code of module "***shift_right***" to verify that the logic obtained is as shown in Figure 4.19.

Example 4.23 Verilog Code for Test Bench to Simulate Module "*shift_right*"

```
module shift_right_tb();

reg [2:0] reg_inputA, reg_inputB;
wire [2:0] wire_outputA;

integer i,j;

initial
begin
    for (i=0; i<8; i=i+1)
        begin
            // to force input stimulus for inputA
            reg_inputA = i;
            for (j=0; j<8; j=j+1)
                begin
                    // to force input stimulus for inputB
                    reg_inputB = j;
                    #10;
                end
        end
end

shift_right shift_right_inst (.inputA(reg_inputA),
.inputB(reg_inputB), .outputA(wire_outputA));

initial
begin
     $monitor ("inputA %b%b%b inputB %b%b%b tempA
%b%b%b outputA %b%b%b",reg_inputA[2], reg_inputA[1],
reg_inputA[0], reg_inputB[2], reg_inputB[1],
reg_inputB[0], shift_right_inst.tempA[2],
shift_right_inst.tempA[1], shift_right_
inst.tempA[0], wire_outputA[2], wire_outputA[1],
wire_outputA[0]);
end

endmodule
```

Example 4.24 shows the simulation results of the test bench module "*shift_right_tb*."

Example 4.24 Simulation Results of Verilog Test Bench Module "*shift_right_tb*"

```
inputA 000 inputB 000 tempA 000 outputA 000
inputA 000 inputB 001 tempA 000 outputA 000
inputA 000 inputB 010 tempA 000 outputA 000
inputA 000 inputB 011 tempA 000 outputA 000
inputA 000 inputB 100 tempA 000 outputA 000
inputA 000 inputB 101 tempA 000 outputA 000
inputA 000 inputB 110 tempA 000 outputA 000
inputA 000 inputB 111 tempA 000 outputA 000
inputA 001 inputB 000 tempA 000 outputA 000
inputA 001 inputB 001 tempA 001 outputA 000
inputA 001 inputB 010 tempA 000 outputA 000
inputA 001 inputB 011 tempA 001 outputA 000
inputA 001 inputB 100 tempA 000 outputA 000
inputA 001 inputB 101 tempA 001 outputA 000
inputA 001 inputB 110 tempA 000 outputA 000
inputA 001 inputB 111 tempA 001 outputA 000
inputA 010 inputB 000 tempA 000 outputA 000
inputA 010 inputB 001 tempA 000 outputA 000
inputA 010 inputB 010 tempA 010 outputA 001
inputA 010 inputB 011 tempA 010 outputA 001
inputA 010 inputB 100 tempA 000 outputA 000
inputA 010 inputB 101 tempA 000 outputA 000
inputA 010 inputB 110 tempA 010 outputA 001
inputA 010 inputB 111 tempA 010 outputA 001
inputA 011 inputB 000 tempA 000 outputA 000
inputA 011 inputB 001 tempA 001 outputA 000
inputA 011 inputB 010 tempA 010 outputA 001
inputA 011 inputB 011 tempA 011 outputA 001
inputA 011 inputB 100 tempA 000 outputA 000
inputA 011 inputB 101 tempA 001 outputA 000
inputA 011 inputB 110 tempA 010 outputA 001
inputA 011 inputB 111 tempA 011 outputA 001
inputA 100 inputB 000 tempA 000 outputA 000
inputA 100 inputB 001 tempA 000 outputA 000
inputA 100 inputB 010 tempA 000 outputA 000
inputA 100 inputB 011 tempA 000 outputA 000
inputA 100 inputB 100 tempA 100 outputA 010
inputA 100 inputB 101 tempA 100 outputA 010
inputA 100 inputB 110 tempA 100 outputA 010
```

```
inputA 100 inputB 111 tempA 100 outputA 010
inputA 101 inputB 000 tempA 000 outputA 000
inputA 101 inputB 001 tempA 001 outputA 000
inputA 101 inputB 010 tempA 000 outputA 000
inputA 101 inputB 011 tempA 001 outputA 000
inputA 101 inputB 100 tempA 100 outputA 010
inputA 101 inputB 101 tempA 101 outputA 010
inputA 101 inputB 110 tempA 100 outputA 010
inputA 101 inputB 111 tempA 101 outputA 010
inputA 110 inputB 000 tempA 000 outputA 000
inputA 110 inputB 001 tempA 000 outputA 000
inputA 110 inputB 010 tempA 010 outputA 001
inputA 110 inputB 011 tempA 010 outputA 001
inputA 110 inputB 100 tempA 100 outputA 010
inputA 110 inputB 101 tempA 100 outputA 010
inputA 110 inputB 110 tempA 110 outputA 011
inputA 110 inputB 111 tempA 110 outputA 011
inputA 111 inputB 000 tempA 000 outputA 000
inputA 111 inputB 001 tempA 001 outputA 000
inputA 111 inputB 010 tempA 010 outputA 001
inputA 111 inputB 011 tempA 011 outputA 001
inputA 111 inputB 100 tempA 100 outputA 010
inputA 111 inputB 101 tempA 101 outputA 010
inputA 111 inputB 110 tempA 110 outputA 011
inputA 111 inputB 111 tempA 111 outputA 011
```

*Note: Notice from the simulation results that the (MSB) is always a zero? This occurs because when shifting right, the MSB is always tagged with logic zero. This causes the synthesized logic for module **shift_right** to have the **outputA(2)** grounded.*

4.8.4 Arithmetic Operator

Verilog allows for five different arithmetic operators that can be used for different operations. They are as follows:

1. addition operator
2. subtraction operator
3. multiplication operator
4. division operator
5. modulus operator

When using these operators, the designer needs to be aware that the logic solution obtained from synthesis may differ if different design constraints are used.

4.8.4.1 Addition operator As the name implies, the addition operator allows an addition operation. It is coded in Verilog by using the symbol "+".

Example 4.25 Verilog Code Using an Addition Operator

```
module addition (inputA, inputB, outputA);

input inputA, inputB;
output [1:0] outputA;

wire [1:0] outputA;

assign outputA = inputA + inputB;

endmodule
```

Figure 4.20 shows a diagram of the synthesized logic module "***addition***" in Example 4.25.

Example 4.26 is a Verilog test bench that can be used to simulate the Verilog code of module "***addition***." The simulation results are shown in Example 4.27.

Example 4.26 Verilog Test Bench to Simulate Verilog Code for Module "*addition*"

```
module addition_tb ();

reg reg_inputA, reg_inputB;
```

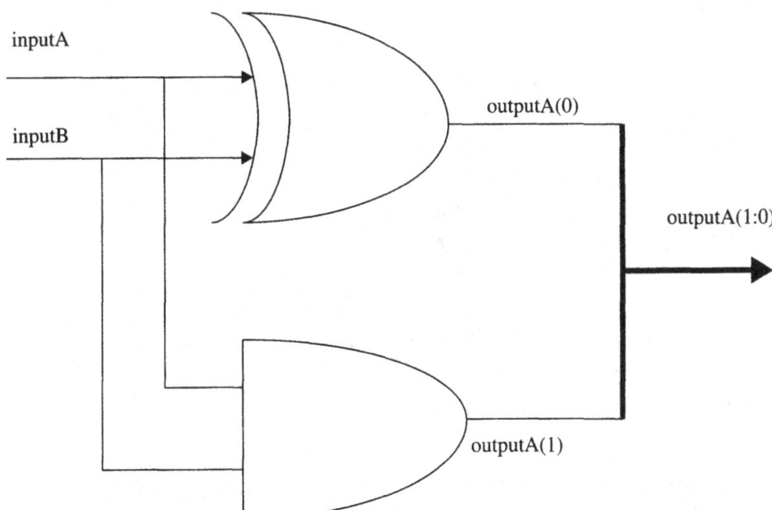

Figure 4.20. Diagram showing synthesized logic for module "addition."

```
wire [1:0] wire_outputA;

integer i,j;

initial
begin
     for (i=0; i<2; i=i+1)
          begin
               reg_inputA = i;
               for (j=0; j<2; j=j+1)
                    begin
                         reg_inputB = j;
                         #10;
                    end
          end
end

addition addition_inst (.inputA(reg_inputA),
.inputB(reg_inputB), .outputA(wire_outputA));

initial
begin
     $monitor ("inputA %b inputB %b outputA %b%b",
reg_inputA, reg_inputB, wire_outputA[1],
wire_outputA[0]);
end

endmodule
```

Example 4.27 Simulation Results for Verilog Test Bench Module "*addition_tb*"

```
inputA 0 inputB 0 outputA 00
inputA 0 inputB 1 outputA 01
inputA 1 inputB 0 outputA 01
inputA 1 inputB 1 outputA 10
```

Appendix A.1 shows the Verilog code for a two-bit by two-bit adder design that uses an addition operator. It also includes a Verilog test bench, simulation results, and the synthesized logic circuit.

4.8.4.2 Subtraction operator As the name implies, the subtraction operator allows a subtract operation. It is coded in Verilog by using the symbol "–".

Example 4.28 Verilog Code Using a Subtraction Operator

```
module subtraction (inputA, inputB, outputA);

input inputA, inputB;
output [1:0] outputA;

wire [1:0] outputA;

assign outputA = inputA - inputB;

endmodule
```

Figure 4.21 shows a diagram of the synthesized logic module "***subtraction***" in Example 4.28.

Example 4.29 is a Verilog test bench that can be used to simulate the Verilog code of module "***subtraction***." The simulation results are shown in Example 4.30.

Example 4.29 Verilog Test Bench to Simulate Verilog Code for Module "*subtraction*"

```
module subtraction_tb ();

reg reg_inputA, reg_inputB;

wire [1:0] wire_outputA;

integer i,j;
```

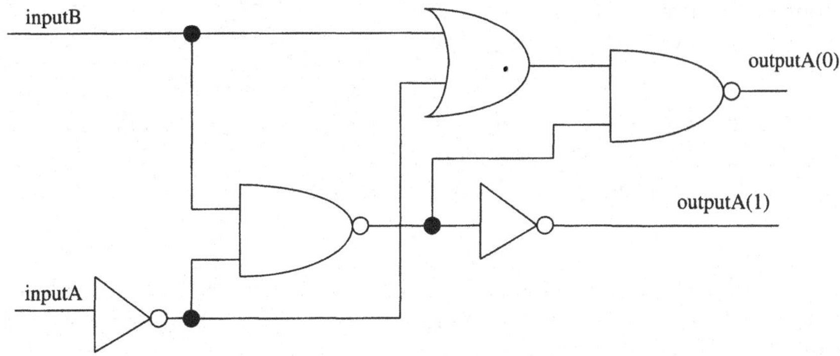

Figure 4.21. Diagram showing synthesized logic for module "subtraction".

```
initial
begin
     for (i=0; i<2; i=i+1)
          begin
               reg_inputA = i;
               for (j=0; j<2; j=j+1)
                    begin
                         reg_inputB = j;
                         #10;
                    end
          end
end

subtraction subtraction_inst (.inputA(reg_inputA),
.inputB(reg_inputB), .outputA(wire_outputA));

initial
begin
     $monitor ("inputA %b inputB %b outputA %b%b",
reg_inputA, reg_inputB, wire_outputA[1],
wire_outputA[0]);
end

endmodule
```

Example 4.30 Simulation Results for Verilog Test Bench Module "*subtraction_tb*"

```
inputA 0 inputB 0 outputA 00
inputA 0 inputB 1 outputA 11
inputA 1 inputB 0 outputA 01
inputA 1 inputB 1 outputA 00
```

When "**inputA** = 0" and "**inputB** = 1", "**inputA** - **inputB** = -1". Two's complement of -1 is "11".

Appendix A.2 shows the Verilog code for a two-bit by two-bit subtractor design that uses a subtraction operator. It also includes a Verilog test bench, simulation results, and the synthesized logic circuit.

4.8.4.3 Multiplication Operator As the name implies, the multiplication operator allows a multiplication operation. It is coded in Verilog by using the symbol "*".

Example 4.31 Verilog Code Using a Multiplication Operator

```
module multiplication (inputA, inputB, outputA);

input [1:0] inputA, inputB;
output [3:0] outputA;

wire [3:0] outputA;

assign outputA = inputA * inputB;

endmodule
```

Figure 4.22 shows a diagram of the synthesized logic module "***multiplication***" in Example 4.31.

Example 4.32 is a Verilog test bench that can be used to simulate the Verilog code of module "***multiplication***." The simulation results are shown in Example 4.33.

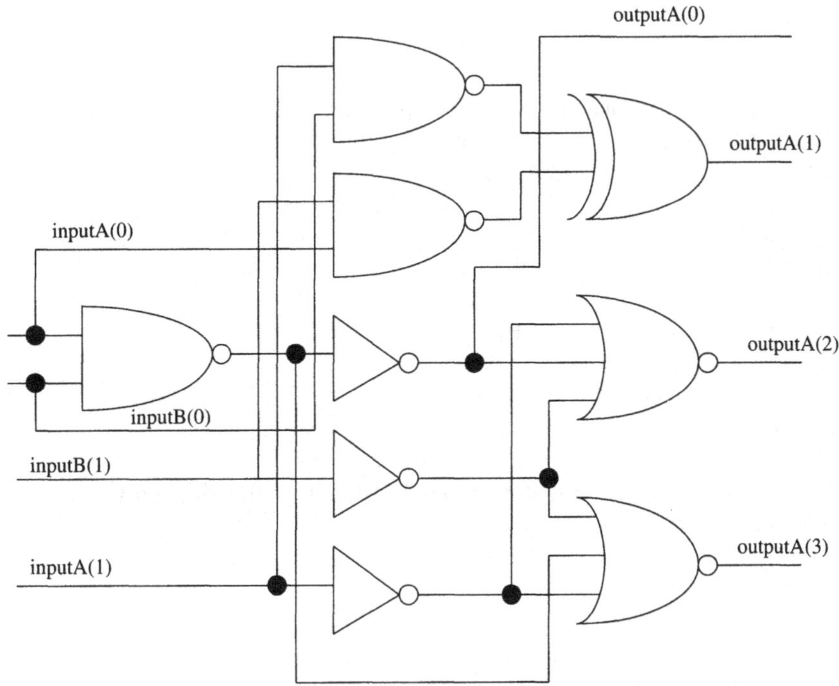

Figure 4.22. Diagram showing synthesized logic for module "multiplication".

Example 4.32 Verilog Test Bench to Simulate Verilog Code for Module "*multiplication*"

```
module multiplication_tb ();

reg [1:0] reg_inputA, reg_inputB;

wire [3:0] wire_outputA;

integer i,j;

initial
begin
     for (i=0; i<4; i=i+1)
          begin
               reg_inputA = i;
               for (j=0; j<4; j=j+1)
                    begin
                         reg_inputB = j;
                         #10;
                    end
          end
end

multiplication multiplication_inst
(.inputA(reg_inputA), .inputB(reg_inputB),
.outputA(wire_outputA));

initial
begin
     $monitor ("inputA %h inputB %h outputA %h",
reg_inputA, reg_inputB, wire_outputA);
end

endmodule
```

Example 4.33 Simulation Results for Verilog Test Bench Module "*multiplication_tb*"

```
inputA 0 inputB 0 outputA 0
inputA 0 inputB 1 outputA 0
inputA 0 inputB 2 outputA 0
inputA 0 inputB 3 outputA 0
inputA 1 inputB 0 outputA 0
inputA 1 inputB 1 outputA 1
```

```
inputA 1 inputB 2 outputA 2
inputA 1 inputB 3 outputA 3
inputA 2 inputB 0 outputA 0
inputA 2 inputB 1 outputA 2
inputA 2 inputB 2 outputA 4
inputA 2 inputB 3 outputA 6
inputA 3 inputB 0 outputA 0
inputA 3 inputB 1 outputA 3
inputA 3 inputB 2 outputA 6
inputA 3 inputB 3 outputA 9
```

Appendix A.3 shows the Verilog code for a four-bit by four-bit multiplier design that uses a multiplication operator. It also includes a Verilog test bench and simulation results.

4.8.5 Division Operator

As the name implies, the division operator allows a division operation. It is coded in Verilog by using the symbol "/". The designer needs to be careful when using the division operator in synthesizable Verilog. The division operator can only be used on constants and not on variables. If the division operator is being used on a value that is not a constant, the synthesis tool will not be able to synthesize the logic.

Example 4.34 Verilog Code Using a Division Operator

```
module division (inputA, inputB, outputA, outputB);

input [3:0] inputA;
input [3:0] inputB;
output [3:0] outputA, outputB;

reg [3:0] outputA, outputB;

always @ (inputA or inputB)
begin
     if (inputA == 4'b1010)
          outputA = 3/3;
     else
          outputA = 0;

     if (inputB == 4'b0011)
          outputB = 8/5;
```

Only constant values can be used when using the division operator in synthesizable Verilog code.

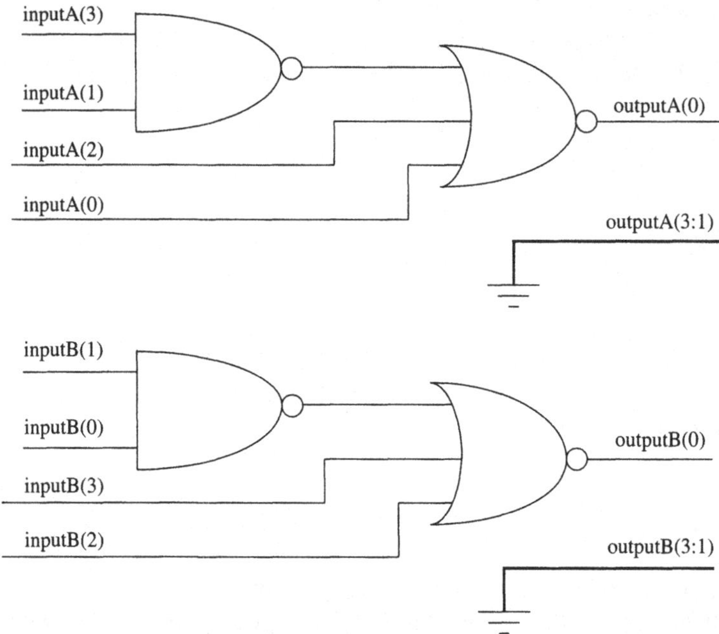

Figure 4.23. Diagram showing synthesized logic for design module "***division***."

```
     else
          outputB = 0;
end

endmodule
```

Figure 4.23 shows a diagram for synthesized logic module "***division***."

Example 4.35 is a Verilog test bench that can be used to simulate the Verilog code of module "***division***". The simulation results are shown in Example 4.36.

Example 4.35 Verilog Test Bench to Simulate Verilog Code for Module "*division*"

```
module division_tb ();

reg [3:0] reg_inputA, reg_inputB;

wire [3:0] wire_outputA, wire_outputB;

integer i,j;

initial
begin
```

```
    for (i=1; i<16; i=i+1)
        begin
            reg_inputA = i;
            for (j=1; j<16; j=j+1)
                begin
                    reg_inputB = j;
                    #10;
                end
        end
end

division division_inst (.inputA(reg_inputA),
.inputB(reg_inputB), .outputA(wire_outputA),
.outputB(wire_outputB));

initial
begin
    $monitor ("inputA %h inputB %h outputA %h outputB
%h", reg_inputA, reg_inputB, wire_outputA,
wire_outputB);
end

endmodule
```

Example 4.36 Simulation Results for Verilog Test Bench Module "*division*"

```
inputA 1 inputB 1 outputA 0 outputB 0
inputA 1 inputB 2 outputA 0 outputB 0
inputA 1 inputB 3 outputA 0 outputB 1
inputA 1 inputB 4 outputA 0 outputB 0
inputA 1 inputB 5 outputA 0 outputB 0
inputA 1 inputB 6 outputA 0 outputB 0
inputA 1 inputB 7 outputA 0 outputB 0
inputA 1 inputB 8 outputA 0 outputB 0
inputA 1 inputB 9 outputA 0 outputB 0
inputA 1 inputB a outputA 0 outputB 0
inputA 1 inputB b outputA 0 outputB 0
inputA 1 inputB c outputA 0 outputB 0
inputA 1 inputB d outputA 0 outputB 0
inputA 1 inputB e outputA 0 outputB 0
inputA 1 inputB f outputA 0 outputB 0
inputA 2 inputB 1 outputA 0 outputB 0
inputA 2 inputB 2 outputA 0 outputB 0
inputA 2 inputB 3 outputA 0 outputB 1
```

When ***inputB*** = "0011" or hexadecimal "3," ***outputB*** is the result of 8/5, which is 1.6. However, in division, any fractions are truncated. Therefore, ***outputB*** would have a value of "1."

```
inputA 2 inputB 4 outputA 0 outputB 0
inputA 2 inputB 5 outputA 0 outputB 0
inputA 2 inputB 6 outputA 0 outputB 0
inputA 2 inputB 7 outputA 0 outputB 0
inputA 2 inputB 8 outputA 0 outputB 0
inputA 2 inputB 9 outputA 0 outputB 0
inputA 2 inputB a outputA 0 outputB 0
inputA 2 inputB b outputA 0 outputB 0
inputA 2 inputB c outputA 0 outputB 0
inputA 2 inputB d outputA 0 outputB 0
inputA 2 inputB e outputA 0 outputB 0
inputA 2 inputB f outputA 0 outputB 0
inputA 3 inputB 1 outputA 0 outputB 0
inputA 3 inputB 2 outputA 0 outputB 0
inputA 3 inputB 3 outputA 0 outputB 1
inputA 3 inputB 4 outputA 0 outputB 0
inputA 3 inputB 5 outputA 0 outputB 0
inputA 3 inputB 6 outputA 0 outputB 0
inputA 3 inputB 7 outputA 0 outputB 0
inputA 3 inputB 8 outputA 0 outputB 0
inputA 3 inputB 9 outputA 0 outputB 0
inputA 3 inputB a outputA 0 outputB 0
inputA 3 inputB b outputA 0 outputB 0
inputA 3 inputB c outputA 0 outputB 0
inputA 3 inputB d outputA 0 outputB 0
inputA 3 inputB e outputA 0 outputB 0
inputA 3 inputB f outputA 0 outputB 0
inputA 4 inputB 1 outputA 0 outputB 0
inputA 4 inputB 2 outputA 0 outputB 0
inputA 4 inputB 3 outputA 0 outputB 1
inputA 4 inputB 4 outputA 0 outputB 0
inputA 4 inputB 5 outputA 0 outputB 0
inputA 4 inputB 6 outputA 0 outputB 0
inputA 4 inputB 7 outputA 0 outputB 0
inputA 4 inputB 8 outputA 0 outputB 0
inputA 4 inputB 9 outputA 0 outputB 0
inputA 4 inputB a outputA 0 outputB 0
inputA 4 inputB b outputA 0 outputB 0
inputA 4 inputB c outputA 0 outputB 0
inputA 4 inputB d outputA 0 outputB 0
inputA 4 inputB e outputA 0 outputB 0
inputA 4 inputB f outputA 0 outputB 0
inputA 5 inputB 1 outputA 0 outputB 0
inputA 5 inputB 2 outputA 0 outputB 0
```

The value of ***inputB*** is "0011." Therefore, ***outputB*** is at logical "1."

```
inputA 5 inputB 3 outputA 0 outputB 1
inputA 5 inputB 4 outputA 0 outputB 0
inputA 5 inputB 5 outputA 0 outputB 0
inputA 5 inputB 6 outputA 0 outputB 0
inputA 5 inputB 7 outputA 0 outputB 0
inputA 5 inputB 8 outputA 0 outputB 0
inputA 5 inputB 9 outputA 0 outputB 0
inputA 5 inputB a outputA 0 outputB 0
inputA 5 inputB b outputA 0 outputB 0
inputA 5 inputB c outputA 0 outputB 0
inputA 5 inputB d outputA 0 outputB 0
inputA 5 inputB e outputA 0 outputB 0
inputA 5 inputB f outputA 0 outputB 0
inputA 6 inputB 1 outputA 0 outputB 0
inputA 6 inputB 2 outputA 0 outputB 0
inputA 6 inputB 3 outputA 0 outputB 1
inputA 6 inputB 4 outputA 0 outputB 0
inputA 6 inputB 5 outputA 0 outputB 0
inputA 6 inputB 6 outputA 0 outputB 0
inputA 6 inputB 7 outputA 0 outputB 0
inputA 6 inputB 8 outputA 0 outputB 0
inputA 6 inputB 9 outputA 0 outputB 0
inputA 6 inputB a outputA 0 outputB 0
inputA 6 inputB b outputA 0 outputB 0
inputA 6 inputB c outputA 0 outputB 0
inputA 6 inputB d outputA 0 outputB 0
inputA 6 inputB e outputA 0 outputB 0
inputA 6 inputB f outputA 0 outputB 0
inputA 7 inputB 1 outputA 0 outputB 0
inputA 7 inputB 2 outputA 0 outputB 0
inputA 7 inputB 3 outputA 0 outputB 1
inputA 7 inputB 4 outputA 0 outputB 0
inputA 7 inputB 5 outputA 0 outputB 0
inputA 7 inputB 6 outputA 0 outputB 0
inputA 7 inputB 7 outputA 0 outputB 0
inputA 7 inputB 8 outputA 0 outputB 0
inputA 7 inputB 9 outputA 0 outputB 0
inputA 7 inputB a outputA 0 outputB 0
inputA 7 inputB b outputA 0 outputB 0
inputA 7 inputB c outputA 0 outputB 0
inputA 7 inputB d outputA 0 outputB 0
inputA 7 inputB e outputA 0 outputB 0
inputA 7 inputB f outputA 0 outputB 0
inputA 8 inputB 1 outputA 0 outputB 0
```

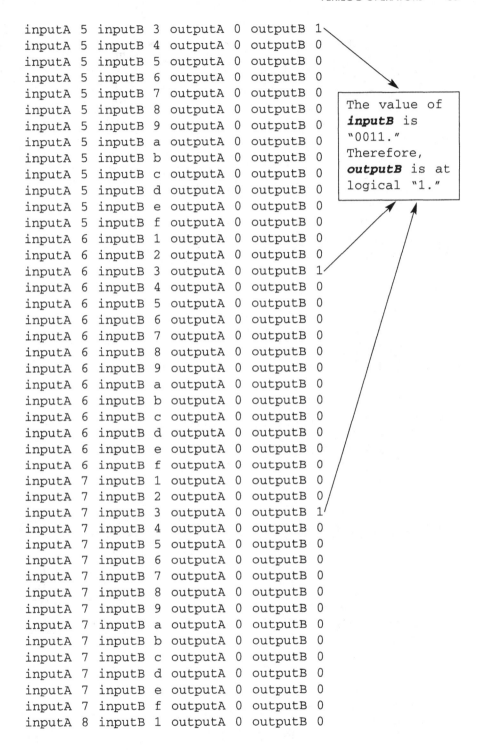

The value of ***inputB*** is "0011." Therefore, ***outputB*** is at logical "1."

```
inputA 8 inputB 2 outputA 0 outputB 0
inputA 8 inputB 3 outputA 0 outputB 1
inputA 8 inputB 4 outputA 0 outputB 0
inputA 8 inputB 5 outputA 0 outputB 0
inputA 8 inputB 6 outputA 0 outputB 0
inputA 8 inputB 7 outputA 0 outputB 0
inputA 8 inputB 8 outputA 0 outputB 0
inputA 8 inputB 9 outputA 0 outputB 0
inputA 8 inputB a outputA 0 outputB 0
inputA 8 inputB b outputA 0 outputB 0
inputA 8 inputB c outputA 0 outputB 0
inputA 8 inputB d outputA 0 outputB 0
inputA 8 inputB e outputA 0 outputB 0
inputA 8 inputB f outputA 0 outputB 0
inputA 9 inputB 1 outputA 0 outputB 0
inputA 9 inputB 2 outputA 0 outputB 0
inputA 9 inputB 3 outputA 0 outputB 1
inputA 9 inputB 4 outputA 0 outputB 0
inputA 9 inputB 5 outputA 0 outputB 0
inputA 9 inputB 6 outputA 0 outputB 0
inputA 9 inputB 7 outputA 0 outputB 0
inputA 9 inputB 8 outputA 0 outputB 0
inputA 9 inputB 9 outputA 0 outputB 0
inputA 9 inputB a outputA 0 outputB 0
inputA 9 inputB b outputA 0 outputB 0
inputA 9 inputB c outputA 0 outputB 0
inputA 9 inputB d outputA 0 outputB 0
inputA 9 inputB e outputA 0 outputB 0
inputA 9 inputB f outputA 0 outputB 0
inputA a inputB 1 outputA 1 outputB 0
inputA a inputB 2 outputA 1 outputB 0
inputA a inputB 3 outputA 1 outputB 1
inputA a inputB 4 outputA 1 outputB 0
inputA a inputB 5 outputA 1 outputB 0
inputA a inputB 6 outputA 1 outputB 0
inputA a inputB 7 outputA 1 outputB 0
inputA a inputB 8 outputA 1 outputB 0
inputA a inputB 9 outputA 1 outputB 0
inputA a inputB a outputA 1 outputB 0
inputA a inputB b outputA 1 outputB 0
inputA a inputB c outputA 1 outputB 0
inputA a inputB d outputA 1 outputB 0
inputA a inputB e outputA 1 outputB 0
inputA a inputB f outputA 1 outputB 0
inputA b inputB 1 outputA 0 outputB 0
```

The value of ***inputB*** is "0011." Therefore, ***outputB*** is at logical "1."

When ***inputA*** = "1010" or hexadecimal "a," ***outputA*** is "1," which is the result of 3/3.

```
inputA b inputB 2 outputA 0 outputB 0
inputA b inputB 3 outputA 0 outputB 1
inputA b inputB 4 outputA 0 outputB 0
inputA b inputB 5 outputA 0 outputB 0
inputA b inputB 6 outputA 0 outputB 0
inputA b inputB 7 outputA 0 outputB 0
inputA b inputB 8 outputA 0 outputB 0
inputA b inputB 9 outputA 0 outputB 0
inputA b inputB a outputA 0 outputB 0
inputA b inputB b outputA 0 outputB 0
inputA b inputB c outputA 0 outputB 0
inputA b inputB d outputA 0 outputB 0
inputA b inputB e outputA 0 outputB 0
inputA b inputB f outputA 0 outputB 0
inputA c inputB 1 outputA 0 outputB 0
inputA c inputB 2 outputA 0 outputB 0
inputA c inputB 3 outputA 0 outputB 1
inputA c inputB 4 outputA 0 outputB 0
inputA c inputB 5 outputA 0 outputB 0
inputA c inputB 6 outputA 0 outputB 0
inputA c inputB 7 outputA 0 outputB 0
inputA c inputB 8 outputA 0 outputB 0
inputA c inputB 9 outputA 0 outputB 0
inputA c inputB a outputA 0 outputB 0
inputA c inputB b outputA 0 outputB 0
inputA c inputB c outputA 0 outputB 0
inputA c inputB d outputA 0 outputB 0
inputA c inputB e outputA 0 outputB 0
inputA c inputB f outputA 0 outputB 0
inputA d inputB 1 outputA 0 outputB 0
inputA d inputB 2 outputA 0 outputB 0
inputA d inputB 3 outputA 0 outputB 1
inputA d inputB 4 outputA 0 outputB 0
inputA d inputB 5 outputA 0 outputB 0
inputA d inputB 6 outputA 0 outputB 0
inputA d inputB 7 outputA 0 outputB 0
inputA d inputB 8 outputA 0 outputB 0
inputA d inputB 9 outputA 0 outputB 0
inputA d inputB a outputA 0 outputB 0
inputA d inputB b outputA 0 outputB 0
inputA d inputB c outputA 0 outputB 0
inputA d inputB d outputA 0 outputB 0
inputA d inputB e outputA 0 outputB 0
inputA d inputB f outputA 0 outputB 0
inputA e inputB 1 outputA 0 outputB 0
```

The value of ***inputB*** is "0011." Therefore, ***outputB*** is at logical "1."

```
inputA e inputB 2 outputA 0 outputB 0
inputA e inputB 3 outputA 0 outputB 1
inputA e inputB 4 outputA 0 outputB 0
inputA e inputB 5 outputA 0 outputB 0
inputA e inputB 6 outputA 0 outputB 0
inputA e inputB 7 outputA 0 outputB 0
inputA e inputB 8 outputA 0 outputB 0
inputA e inputB 9 outputA 0 outputB 0
inputA e inputB a outputA 0 outputB 0
inputA e inputB b outputA 0 outputB 0
inputA e inputB c outputA 0 outputB 0
inputA e inputB d outputA 0 outputB 0
inputA e inputB e outputA 0 outputB 0
inputA e inputB f outputA 0 outputB 0
inputA f inputB 1 outputA 0 outputB 0
inputA f inputB 2 outputA 0 outputB 0
inputA f inputB 3 outputA 0 outputB 1
inputA f inputB 4 outputA 0 outputB 0
inputA f inputB 5 outputA 0 outputB 0
inputA f inputB 6 outputA 0 outputB 0
inputA f inputB 7 outputA 0 outputB 0
inputA f inputB 8 outputA 0 outputB 0
inputA f inputB 9 outputA 0 outputB 0
inputA f inputB a outputA 0 outputB 0
inputA f inputB b outputA 0 outputB 0
inputA f inputB c outputA 0 outputB 0
inputA f inputB d outputA 0 outputB 0
inputA f inputB e outputA 0 outputB 0
inputA f inputB f outputA 0 outputB 0
```

Note: Only constant values can be used when using the division operator in synthesizable Verilog code. Values obtained using the division operator are in integer format and do not have any fractions.

4.8.6 Modulus Operator

Modulus operator allows an arithmetic operation that returns a value of the remainder of a division (the operator is coded in Verilog by using the symbol "%"). For example, a modulus operation of "5 % 3" would return a value of 2, which is the remainder of the division operation of "5/3".

The designer needs to be careful when using the modulus operator in synthesizable Verilog. The operator can only be used on constants and not on variables. If the modulus operator is being used on a value that is not a constant, the synthesis tool will not be able to synthesize the logic.

Example 4.37 Verilog Code Using a Modulus Operator

```
module modulus (inputA, inputB, outputA, outputB);

input [3:0] inputA;
input [3:0] inputB;
output [3:0] outputA, outputB;

reg [3:0] outputA, outputB;

always @ (inputA or inputB)
begin
     if (inputA == 4'b1010)
          outputA = 2 % 5;
     else
          outputA = 0;

     if (inputB == 4'b0011)
          outputB = 8 % 5;
     else
          outputB = 0;
end

endmodule
```

Figure 4.24 shows a diagram of the synthesized logic module "***modulus***."

Example 4.38 is a verilog test bench that can be used to simulate the verilog code of module "***modulus***". The simulation results are shown in Example 4.39.

Example 4.38 Verilog Test Bench to Simulate Verilog Code for Module "*modulus*"

```
module modulus_tb ();

reg [3:0] reg_inputA, reg_inputB;

wire [3:0] wire_outputA, wire_outputB;

integer i,j;

initial
begin
     for (i=1; i<16; i=i+1)
          begin
               reg_inputA = i;
```

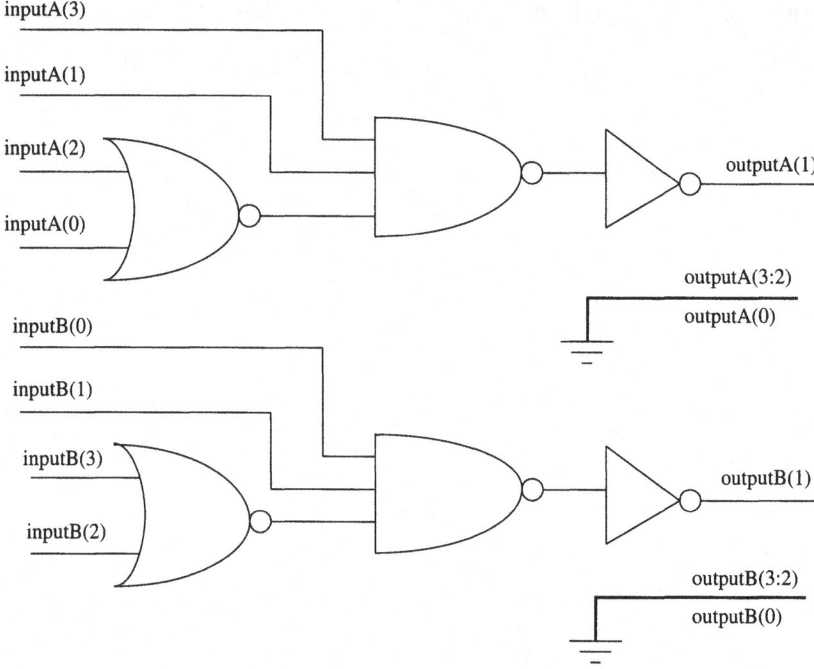

Figure 4.24. Diagram showing synthesized logic for design module "***modulus***."

```
                for (j=1; j<16; j=j+1)
                     begin
                          reg_inputB = j;
                          #10;
                     end
          end
end

modulus modulus_inst (.inputA(reg_inputA),
.inputB(reg_inputB), .outputA(wire_outputA),
.outputB(wire_outputB));

initial
begin
     $monitor ("inputA %h inputB %h outputA %h outputB
%h", reg_inputA, reg_inputB, wire_outputA,
wire_outputB);
end

endmodule
```

Example 4.39 Simulation Results for Verilog Test Bench Module "*modulus*"

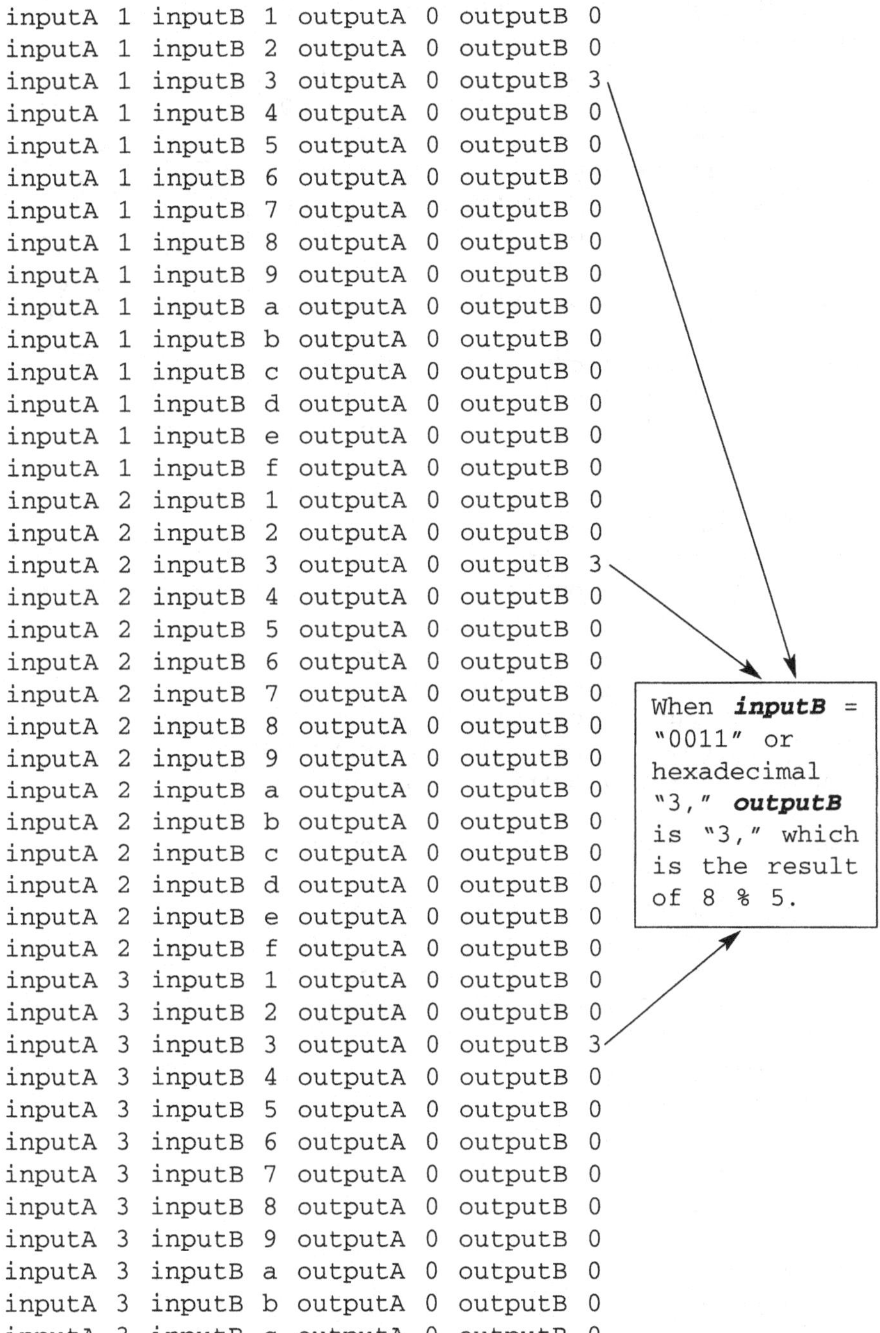

```
inputA 1 inputB 1 outputA 0 outputB 0
inputA 1 inputB 2 outputA 0 outputB 0
inputA 1 inputB 3 outputA 0 outputB 3
inputA 1 inputB 4 outputA 0 outputB 0
inputA 1 inputB 5 outputA 0 outputB 0
inputA 1 inputB 6 outputA 0 outputB 0
inputA 1 inputB 7 outputA 0 outputB 0
inputA 1 inputB 8 outputA 0 outputB 0
inputA 1 inputB 9 outputA 0 outputB 0
inputA 1 inputB a outputA 0 outputB 0
inputA 1 inputB b outputA 0 outputB 0
inputA 1 inputB c outputA 0 outputB 0
inputA 1 inputB d outputA 0 outputB 0
inputA 1 inputB e outputA 0 outputB 0
inputA 1 inputB f outputA 0 outputB 0
inputA 2 inputB 1 outputA 0 outputB 0
inputA 2 inputB 2 outputA 0 outputB 0
inputA 2 inputB 3 outputA 0 outputB 3
inputA 2 inputB 4 outputA 0 outputB 0
inputA 2 inputB 5 outputA 0 outputB 0
inputA 2 inputB 6 outputA 0 outputB 0
inputA 2 inputB 7 outputA 0 outputB 0
inputA 2 inputB 8 outputA 0 outputB 0
inputA 2 inputB 9 outputA 0 outputB 0
inputA 2 inputB a outputA 0 outputB 0
inputA 2 inputB b outputA 0 outputB 0
inputA 2 inputB c outputA 0 outputB 0
inputA 2 inputB d outputA 0 outputB 0
inputA 2 inputB e outputA 0 outputB 0
inputA 2 inputB f outputA 0 outputB 0
inputA 3 inputB 1 outputA 0 outputB 0
inputA 3 inputB 2 outputA 0 outputB 0
inputA 3 inputB 3 outputA 0 outputB 3
inputA 3 inputB 4 outputA 0 outputB 0
inputA 3 inputB 5 outputA 0 outputB 0
inputA 3 inputB 6 outputA 0 outputB 0
inputA 3 inputB 7 outputA 0 outputB 0
inputA 3 inputB 8 outputA 0 outputB 0
inputA 3 inputB 9 outputA 0 outputB 0
inputA 3 inputB a outputA 0 outputB 0
inputA 3 inputB b outputA 0 outputB 0
inputA 3 inputB c outputA 0 outputB 0
```

When inputB = "0011" or hexadecimal "3," outputB is "3," which is the result of 8 % 5.

```
inputA 3 inputB d outputA 0 outputB 0
inputA 3 inputB e outputA 0 outputB 0
inputA 3 inputB f outputA 0 outputB 0
inputA 4 inputB 1 outputA 0 outputB 0
inputA 4 inputB 2 outputA 0 outputB 0
inputA 4 inputB 3 outputA 0 outputB 3
inputA 4 inputB 4 outputA 0 outputB 0
inputA 4 inputB 5 outputA 0 outputB 0
inputA 4 inputB 6 outputA 0 outputB 0
inputA 4 inputB 7 outputA 0 outputB 0
inputA 4 inputB 8 outputA 0 outputB 0
inputA 4 inputB 9 outputA 0 outputB 0
inputA 4 inputB a outputA 0 outputB 0
inputA 4 inputB b outputA 0 outputB 0
inputA 4 inputB c outputA 0 outputB 0
inputA 4 inputB d outputA 0 outputB 0
inputA 4 inputB e outputA 0 outputB 0
inputA 4 inputB f outputA 0 outputB 0
inputA 5 inputB 1 outputA 0 outputB 0
inputA 5 inputB 2 outputA 0 outputB 0
inputA 5 inputB 3 outputA 0 outputB 3
inputA 5 inputB 4 outputA 0 outputB 0
inputA 5 inputB 5 outputA 0 outputB 0
inputA 5 inputB 6 outputA 0 outputB 0
inputA 5 inputB 7 outputA 0 outputB 0
inputA 5 inputB 8 outputA 0 outputB 0
inputA 5 inputB 9 outputA 0 outputB 0
inputA 5 inputB a outputA 0 outputB 0
inputA 5 inputB b outputA 0 outputB 0
inputA 5 inputB c outputA 0 outputB 0
inputA 5 inputB d outputA 0 outputB 0
inputA 5 inputB e outputA 0 outputB 0
inputA 5 inputB f outputA 0 outputB 0
inputA 6 inputB 1 outputA 0 outputB 0
inputA 6 inputB 2 outputA 0 outputB 0
inputA 6 inputB 3 outputA 0 outputB 3
inputA 6 inputB 4 outputA 0 outputB 0
inputA 6 inputB 5 outputA 0 outputB 0
inputA 6 inputB 6 outputA 0 outputB 0
inputA 6 inputB 7 outputA 0 outputB 0
inputA 6 inputB 8 outputA 0 outputB 0
inputA 6 inputB 9 outputA 0 outputB 0
inputA 6 inputB a outputA 0 outputB 0
inputA 6 inputB b outputA 0 outputB 0
inputA 6 inputB c outputA 0 outputB 0
```

When ***inputB*** = "0011" or hexadecimal "3," ***outputB*** is "3," which is the result of 8 % 5.

```
inputA 6 inputB d outputA 0 outputB 0
inputA 6 inputB e outputA 0 outputB 0
inputA 6 inputB f outputA 0 outputB 0
inputA 7 inputB 1 outputA 0 outputB 0
inputA 7 inputB 2 outputA 0 outputB 0
inputA 7 inputB 3 outputA 0 outputB 3
inputA 7 inputB 4 outputA 0 outputB 0
inputA 7 inputB 5 outputA 0 outputB 0
inputA 7 inputB 6 outputA 0 outputB 0
inputA 7 inputB 7 outputA 0 outputB 0
inputA 7 inputB 8 outputA 0 outputB 0
inputA 7 inputB 9 outputA 0 outputB 0
inputA 7 inputB a outputA 0 outputB 0
inputA 7 inputB b outputA 0 outputB 0
inputA 7 inputB c outputA 0 outputB 0
inputA 7 inputB d outputA 0 outputB 0
inputA 7 inputB e outputA 0 outputB 0
inputA 7 inputB f outputA 0 outputB 0
inputA 8 inputB 1 outputA 0 outputB 0
inputA 8 inputB 2 outputA 0 outputB 0
inputA 8 inputB 3 outputA 0 outputB 3
inputA 8 inputB 4 outputA 0 outputB 0
inputA 8 inputB 5 outputA 0 outputB 0
inputA 8 inputB 6 outputA 0 outputB 0
inputA 8 inputB 7 outputA 0 outputB 0
inputA 8 inputB 8 outputA 0 outputB 0
inputA 8 inputB 9 outputA 0 outputB 0
inputA 8 inputB a outputA 0 outputB 0
inputA 8 inputB b outputA 0 outputB 0
inputA 8 inputB c outputA 0 outputB 0
inputA 8 inputB d outputA 0 outputB 0
inputA 8 inputB e outputA 0 outputB 0
inputA 8 inputB f outputA 0 outputB 0
inputA 9 inputB 1 outputA 0 outputB 0
inputA 9 inputB 2 outputA 0 outputB 0
inputA 9 inputB 3 outputA 0 outputB 3
inputA 9 inputB 4 outputA 0 outputB 0
inputA 9 inputB 5 outputA 0 outputB 0
inputA 9 inputB 6 outputA 0 outputB 0
inputA 9 inputB 7 outputA 0 outputB 0
inputA 9 inputB 8 outputA 0 outputB 0
inputA 9 inputB 9 outputA 0 outputB 0
inputA 9 inputB a outputA 0 outputB 0
inputA 9 inputB b outputA 0 outputB 0
inputA 9 inputB c outputA 0 outputB 0
```

When ***inputB*** = "0011" or hexadecimal "3," ***outputB*** is "3," which is the result of 8 % 5.

```
inputA 9 inputB d outputA 0 outputB 0
inputA 9 inputB e outputA 0 outputB 0
inputA 9 inputB f outputA 0 outputB 0
inputA a inputB 1 outputA 2 outputB 0
inputA a inputB 2 outputA 2 outputB 0
inputA a inputB 3 outputA 2 outputB 3
inputA a inputB 4 outputA 2 outputB 0
inputA a inputB 5 outputA 2 outputB 0
inputA a inputB 6 outputA 2 outputB 0
inputA a inputB 7 outputA 2 outputB 0
inputA a inputB 8 outputA 2 outputB 0
inputA a inputB 9 outputA 2 outputB 0
inputA a inputB a outputA 2 outputB 0
inputA a inputB b outputA 2 outputB 0
inputA a inputB c outputA 2 outputB 0
inputA a inputB d outputA 2 outputB 0
inputA a inputB e outputA 2 outputB 0
inputA a inputB f outputA 2 outputB 0
inputA b inputB 1 outputA 0 outputB 0
inputA b inputB 2 outputA 0 outputB 0
inputA b inputB 3 outputA 0 outputB 3
inputA b inputB 4 outputA 0 outputB 0
inputA b inputB 5 outputA 0 outputB 0
inputA b inputB 6 outputA 0 outputB 0
inputA b inputB 7 outputA 0 outputB 0
inputA b inputB 8 outputA 0 outputB 0
inputA b inputB 9 outputA 0 outputB 0
inputA b inputB a outputA 0 outputB 0
inputA b inputB b outputA 0 outputB 0
inputA b inputB c outputA 0 outputB 0
inputA b inputB d outputA 0 outputB 0
inputA b inputB e outputA 0 outputB 0
inputA b inputB f outputA 0 outputB 0
inputA c inputB 1 outputA 0 outputB 0
inputA c inputB 2 outputA 0 outputB 0
inputA c inputB 3 outputA 0 outputB 3
inputA c inputB 4 outputA 0 outputB 0
inputA c inputB 5 outputA 0 outputB 0
inputA c inputB 6 outputA 0 outputB 0
inputA c inputB 7 outputA 0 outputB 0
inputA c inputB 8 outputA 0 outputB 0
inputA c inputB 9 outputA 0 outputB 0
inputA c inputB a outputA 0 outputB 0
inputA c inputB b outputA 0 outputB 0
inputA c inputB c outputA 0 outputB 0
```

When ***inputA*** = "1010" or hexadecimal "a," ***outputA*** is "2," which is the result of 2 % 5.

When ***inputB*** = "0011" or hexadecimal "3," ***outputB*** is "3," which is the result of 8 % 5.

When ***inputB*** = "0011" or hexadecimal "3," ***outputB*** is "3," which is the result of 8 % 5.

```
inputA c inputB d outputA 0 outputB 0
inputA c inputB e outputA 0 outputB 0
inputA c inputB f outputA 0 outputB 0
inputA d inputB 1 outputA 0 outputB 0
inputA d inputB 2 outputA 0 outputB 0
inputA d inputB 3 outputA 0 outputB 3
inputA d inputB 4 outputA 0 outputB 0
inputA d inputB 5 outputA 0 outputB 0
inputA d inputB 6 outputA 0 outputB 0
inputA d inputB 7 outputA 0 outputB 0
inputA d inputB 8 outputA 0 outputB 0
inputA d inputB 9 outputA 0 outputB 0
inputA d inputB a outputA 0 outputB 0
inputA d inputB b outputA 0 outputB 0
inputA d inputB c outputA 0 outputB 0
inputA d inputB d outputA 0 outputB 0
inputA d inputB e outputA 0 outputB 0
inputA d inputB f outputA 0 outputB 0
inputA e inputB 1 outputA 0 outputB 0
inputA e inputB 2 outputA 0 outputB 0
inputA e inputB 3 outputA 0 outputB 3
inputA e inputB 4 outputA 0 outputB 0
inputA e inputB 5 outputA 0 outputB 0
inputA e inputB 6 outputA 0 outputB 0
inputA e inputB 7 outputA 0 outputB 0
inputA e inputB 8 outputA 0 outputB 0
inputA e inputB 9 outputA 0 outputB 0
inputA e inputB a outputA 0 outputB 0
inputA e inputB b outputA 0 outputB 0
inputA e inputB c outputA 0 outputB 0
inputA e inputB d outputA 0 outputB 0
inputA e inputB e outputA 0 outputB 0
inputA e inputB f outputA 0 outputB 0
inputA f inputB 1 outputA 0 outputB 0
inputA f inputB 2 outputA 0 outputB 0
inputA f inputB 3 outputA 0 outputB 3
inputA f inputB 4 outputA 0 outputB 0
inputA f inputB 5 outputA 0 outputB 0
inputA f inputB 6 outputA 0 outputB 0
inputA f inputB 7 outputA 0 outputB 0
inputA f inputB 8 outputA 0 outputB 0
inputA f inputB 9 outputA 0 outputB 0
inputA f inputB a outputA 0 outputB 0
inputA f inputB b outputA 0 outputB 0
inputA f inputB c outputA 0 outputB 0
```

```
inputA f inputB d outputA 0 outputB 0
inputA f inputB e outputA 0 outputB 0
inputA f inputB f outputA 0 outputB 0
```

4.8.7 Logical Operator

Logical operators operate on a group of operands and return the result of the operation as a single-bit result of either 1 or 0. The operands can be single bit or multiple bit, but the result of the operation is always in single bit of 1 (true condition) or 0 (false condition). There are three different logical operators that can be used in Verilog:

1. && This is a logical-AND operator. It performs an AND function on the operands to return a single-bit value.
2. || This is a logical-OR operator. It performs an OR function on the operands to return a single-bit value.
3. ! This is a logical-NOT operator. It performs an inversion (NOT function) on the operand to return a single-bit value.

Example 4.40 shows a Verilog code that uses logical operators. The diagram in Figure 4.25 shows the synthesized logic for the Verilog code of module "***logical***" in Example 4.40.

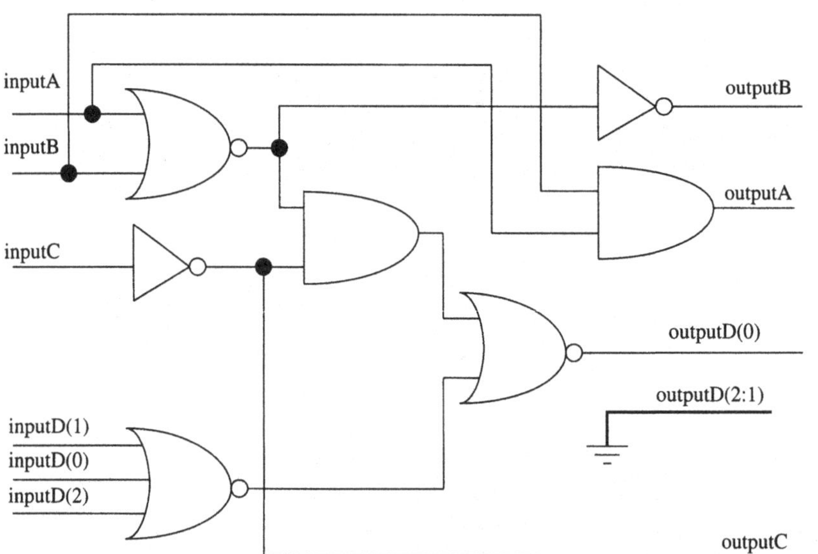

FIGURE 4.25. Diagram showing synthesized logic for verilog code module "logical."

Example 4.40 Verilog Code Using Logical Operators

```
module logical (inputA, inputB, inputC, inputD,
outputA, outputB, outputC, outputD);

input inputA, inputB, inputC;
input [2:0] inputD;

output outputA, outputB, outputC;
output [2:0] outputD;

// for logical AND
assign outputA = inputA && inputB;

// for logical OR
assign outputB = inputA || inputB;

// for logical NOT
assign outputC = !inputC;

// for vector format

assign outputD = {inputA, inputB, inputC} && inputD;

endmodule
```

```
three-bit bus operands
```

```
Although operands are three bits,
the result of the logical operation
&& remains only one-bit wide with
the upper two bits being grounded.
```

Referring to Figure 4.25, notice how the output bus ***outputD(2:0)*** has bits (2:1) grounded whereas bit (0) is connected to logical gates. The ***outputD(2:1)*** is grounded because the logical operator returns a result that is only one bit wide, which represents either a true (logic 1) or false (logic 0) condition. Although ***inputD*** is a three-bit bus operand,

```
assign outputD = {inputA, inputB, inputC} && inputD;
```

concatenation of ***inputA***, ***inputB***, and ***inputC*** is also a three-bit bus, the result of ***outputD*** is only a single bit. The two upper bits of 1 and 2 are grounded.

Example 4.41 is a Verilog test bench that can be used to simulate the design module "***logical***" in Example 4.40. Example 4.42 shows the simulation results from the test bench.

Example 4.41 Verilog Test Bench for Simulating Module "*logical*"

```
module logical_tb ();

reg inputA, inputB, inputC;
reg [2:0] inputD;

integer i,j;

initial
begin
     for (i=0; i<8; i=i+1)
          begin
               {inputA, inputB, inputC} = i;
               for (j=0; j<8; j=j+1)
                    begin
                         inputD = j;
                         #10;
                    end
          end
end

logical logical_inst (.inputA(inputA), .inputB(inputB),
.inputC(inputC), .inputD(inputD), .outputA(outputA),
.outputB(outputB), .outputC(outputC),
.outputD(outputD));

initial
begin
     $monitor ("inputA %b inputB %b inputC %b inputD
%h outputA %b outputB %b outputC %b outputD
%h",inputA, inputB, inputC, inputD, outputA, outputB,
outputC, outputD);
end

endmodule
```

Example 4.42 Simulation Results for Verilog Test Bench Module "*logical_tb*"

```
inputA 0 inputB 0 inputC 0 inputD 0 outputA 0 outputB
0 outputC 1 outputD 0
inputA 0 inputB 0 inputC 0 inputD 1 outputA 0 outputB
0 outputC 1 outputD 0
```

```
inputA 0 inputB 0 inputC 0 inputD 2 outputA 0 outputB
0 outputC 1 outputD 0
inputA 0 inputB 0 inputC 0 inputD 3 outputA 0 outputB
0 outputC 1 outputD 0
inputA 0 inputB 0 inputC 0 inputD 4 outputA 0 outputB
0 outputC 1 outputD 0
inputA 0 inputB 0 inputC 0 inputD 5 outputA 0 outputB
0 outputC 1 outputD 0
inputA 0 inputB 0 inputC 0 inputD 6 outputA 0 outputB
0 outputC 1 outputD 0
inputA 0 inputB 0 inputC 0 inputD 7 outputA 0 outputB
0 outputC 1 outputD 0
inputA 0 inputB 0 inputC 1 inputD 0 outputA 0 outputB
0 outputC 0 outputD 0
inputA 0 inputB 0 inputC 1 inputD 1 outputA 0 outputB
0 outputC 0 outputD 1
inputA 0 inputB 0 inputC 1 inputD 2 outputA 0 outputB
0 outputC 0 outputD 1
inputA 0 inputB 0 inputC 1 inputD 3 outputA 0 outputB
0 outputC 0 outputD 1
inputA 0 inputB 0 inputC 1 inputD 4 outputA 0 outputB
0 outputC 0 outputD 1
inputA 0 inputB 0 inputC 1 inputD 5 outputA 0 outputB
0 outputC 0 outputD 1
inputA 0 inputB 0 inputC 1 inputD 6 outputA 0 outputB
0 outputC 0 outputD 1
inputA 0 inputB 0 inputC 1 inputD 7 outputA 0 outputB
0 outputC 0 outputD 1
inputA 0 inputB 1 inputC 0 inputD 0 outputA 0 outputB
1 outputC 1 outputD 0
inputA 0 inputB 1 inputC 0 inputD 1 outputA 0 outputB
1 outputC 1 outputD 1
inputA 0 inputB 1 inputC 0 inputD 2 outputA 0 outputB
1 outputC 1 outputD 1
inputA 0 inputB 1 inputC 0 inputD 3 outputA 0 outputB
1 outputC 1 outputD 1
inputA 0 inputB 1 inputC 0 inputD 4 outputA 0 outputB
1 outputC 1 outputD 1
inputA 0 inputB 1 inputC 0 inputD 5 outputA 0 outputB
1 outputC 1 outputD 1
inputA 0 inputB 1 inputC 0 inputD 6 outputA 0 outputB
1 outputC 1 outputD 1
inputA 0 inputB 1 inputC 0 inputD 7 outputA 0 outputB
1 outputC 1 outputD 1
```

```
inputA 0 inputB 1 inputC 1 inputD 0 outputA 0 outputB
1 outputC 0 outputD 0
inputA 0 inputB 1 inputC 1 inputD 1 outputA 0 outputB
1 outputC 0 outputD 1
inputA 0 inputB 1 inputC 1 inputD 2 outputA 0 outputB
1 outputC 0 outputD 1
inputA 0 inputB 1 inputC 1 inputD 3 outputA 0 outputB
1 outputC 0 outputD 1
inputA 0 inputB 1 inputC 1 inputD 4 outputA 0 outputB
1 outputC 0 outputD 1
inputA 0 inputB 1 inputC 1 inputD 5 outputA 0 outputB
1 outputC 0 outputD 1
inputA 0 inputB 1 inputC 1 inputD 6 outputA 0 outputB
1 outputC 0 outputD 1
inputA 0 inputB 1 inputC 1 inputD 7 outputA 0 outputB
1 outputC 0 outputD 1
inputA 1 inputB 0 inputC 0 inputD 0 outputA 0 outputB
1 outputC 1 outputD 0
inputA 1 inputB 0 inputC 0 inputD 1 outputA 0 outputB
1 outputC 1 outputD 1
inputA 1 inputB 0 inputC 0 inputD 2 outputA 0 outputB
1 outputC 1 outputD 1
inputA 1 inputB 0 inputC 0 inputD 3 outputA 0 outputB
1 outputC 1 outputD 1
inputA 1 inputB 0 inputC 0 inputD 4 outputA 0 outputB
1 outputC 1 outputD 1
inputA 1 inputB 0 inputC 0 inputD 5 outputA 0 outputB
1 outputC 1 outputD 1
inputA 1 inputB 0 inputC 0 inputD 6 outputA 0 outputB
1 outputC 1 outputD 1
inputA 1 inputB 0 inputC 0 inputD 7 outputA 0 outputB
1 outputC 1 outputD 1
inputA 1 inputB 0 inputC 1 inputD 0 outputA 0 outputB
1 outputC 0 outputD 0
inputA 1 inputB 0 inputC 1 inputD 1 outputA 0 outputB
1 outputC 0 outputD 1
inputA 1 inputB 0 inputC 1 inputD 2 outputA 0 outputB
1 outputC 0 outputD 1
inputA 1 inputB 0 inputC 1 inputD 3 outputA 0 outputB
1 outputC 0 outputD 1
inputA 1 inputB 0 inputC 1 inputD 4 outputA 0 outputB
1 outputC 0 outputD 1
inputA 1 inputB 0 inputC 1 inputD 5 outputA 0 outputB
1 outputC 0 outputD 1
inputA 1 inputB 0 inputC 1 inputD 6 outputA 0 outputB
1 outputC 0 outputD 1
```

```
inputA 1 inputB 0 inputC 1 inputD 7 outputA 0 outputB
1 outputC 0 outputD 1
inputA 1 inputB 1 inputC 0 inputD 0 outputA 1 outputB
1 outputC 1 outputD 0
inputA 1 inputB 1 inputC 0 inputD 1 outputA 1 outputB
1 outputC 1 outputD 1
inputA 1 inputB 1 inputC 0 inputD 2 outputA 1 outputB
1 outputC 1 outputD 1
inputA 1 inputB 1 inputC 0 inputD 3 outputA 1 outputB
1 outputC 1 outputD 1
inputA 1 inputB 1 inputC 0 inputD 4 outputA 1 outputB
1 outputC 1 outputD 1
inputA 1 inputB 1 inputC 0 inputD 5 outputA 1 outputB
1 outputC 1 outputD 1
inputA 1 inputB 1 inputC 0 inputD 6 outputA 1 outputB
1 outputC 1 outputD 1
inputA 1 inputB 1 inputC 0 inputD 7 outputA 1 outputB
1 outputC 1 outputD 1
inputA 1 inputB 1 inputC 1 inputD 0 outputA 1 outputB
1 outputC 0 outputD 0
inputA 1 inputB 1 inputC 1 inputD 1 outputA 1 outputB
1 outputC 0 outputD 1
inputA 1 inputB 1 inputC 1 inputD 2 outputA 1 outputB
1 outputC 0 outputD 1
inputA 1 inputB 1 inputC 1 inputD 3 outputA 1 outputB
1 outputC 0 outputD 1
inputA 1 inputB 1 inputC 1 inputD 4 outputA 1 outputB
1 outputC 0 outputD 1
inputA 1 inputB 1 inputC 1 inputD 5 outputA 1 outputB
1 outputC 0 outputD 1
inputA 1 inputB 1 inputC 1 inputD 6 outputA 1 outputB
1 outputC 0 outputD 1
inputA 1 inputB 1 inputC 1 inputD 7 outputA 1 outputB
1 outputC 0 outputD 1
```

Referring to the simulation results shown in Example 4.42, note that the bus assignment to create ***outputD*** (*assign outputD = {inputA, inputB, inputC} && inputD;*) has a logical function that has ***outputD(0)*** at a logical 1 when any of ***inputA***, ***inputB***, ***inputC*** is a logical 1 ***AND*** any of the three bits of ***inputD*** is a logical 1.

4.8.8 Bitwise Operator

Bitwise operators are similar to logical operators except that bitwise operators operate on buses and return the logic result in bus form. For example, if a bitwise operator is used on two three-bit operands, the result of the operation would be a three-bit result. There are four types of bitwise operators:

1. & This is a bitwise-AND operator. It performs an AND function on the operands to return a value that is equivalent in bus width to the operands.
2. | This is a bitwise-OR operator. It performs an OR function on the operands to return a value that is equivalent in bus width to the operands.
3. ~ This is a bitwise-NOT operator. It performs an inversion (NOT function) on the operand to return a value that is equivalent in bus width to the operands.
4. ∧ This is a bitwise-XOR operator. It performs an XOR function on the operands to return a value that is equivalent in bus width to the operands.

Example 4.43 shows a Verilog code that uses bitwise operators. The diagram in Figure 4.26 shows the synthesized logic for the Verilog code of Example 4.43.

Example 4.43 Verilog Code Using Bitwise Operators

```
module bitwise (inputA, inputB, inputC, inputD,
outputA, outputB, outputC, outputD, outputE);

input inputA, inputB, inputC;
input [2:0] inputD;

output outputA, outputB, outputC, outputE;
output [2:0] outputD;

wire outputA, outputB, outputC, outputE;
wire [2:0] outputD;

// for bitwise AND
assign outputA = inputA & inputB;

// for bitwise OR
assign outputB = inputA | inputB;

// for bitwise NOT
assign outputC = ~inputC;

// for bitwise XOR
assign outputE = inputA ^ inputB;
```

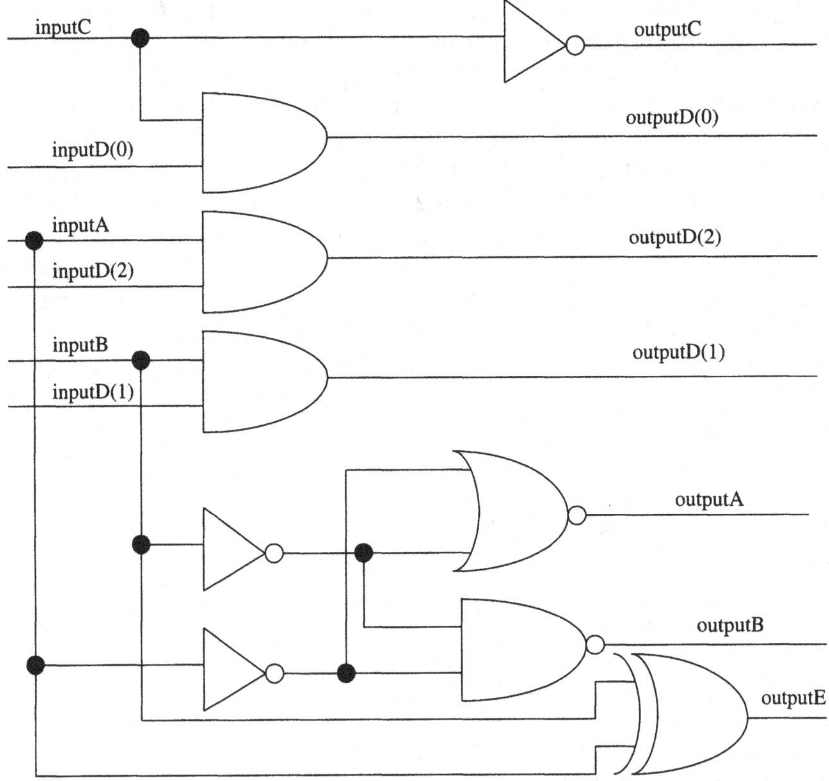

FIGURE 4.26. Diagram showing synthesized logic for verilog code module "*bitwise*."

```
// for vector format
assign outputD = {inputA, inputB, inputC} & inputD;

endmodule
```

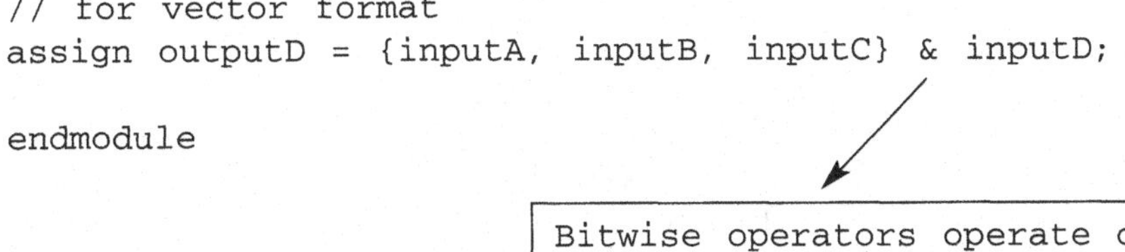

```
Bitwise operators operate on
the operands bit by bit and
return a result that has the
same bus width.
```

Notice how the Verilog code of module "***logical***" in Example 4.40 that uses logical operators (&&, ||, and ! operators) are somewhat similar to the Verilog code of module "***bitwise***" in Example 4.43 that uses bitwise operators (&, |, and ~ operators)? Although both the Verilog codes are identical (except the operators), both would simulate and synthesize to relatively different results for ***outputD(2:0)***. The functionality for ***outputA***, ***outputB***, and ***outputC*** remains the same for both Verilog codes in Examples 4.40 and Example 4.43.

Note: Logical operators (Example 4.40) do not have XOR function. Only bitwise operators (Example 4.43) have XOR function. To obtain the bitwise operator of XNOR, the symbol ~^ or ^~ can be used.

Referring to Figure 4.26, ***outputD(2:1)*** is not grounded, as is the case of the synthesized logic for logical operator shown in Figure 4.25. Therefore, it is important for the designer to note that use of a bitwise operator *(assign outputD = {inputA, inputB, inputC} & inputD;)* would create a result that has the same bus width as the operands.

Example 4.44 is a Verilog test bench that can be used to simulate the Verilog code of module "***bitwise***" in Example 4.43. Example 4.45 shows the simulation results.

Example 4.44 Verilog Test Bench for Simulation of Verilog Code Module "*bitwise*"

```
module bitwise_tb ();

reg inputA, inputB, inputC;
reg [2:0] inputD;
wire outputA, outputB, outputC;
wire [2:0] outputD;

integer i,j;

initial
begin
     for (i=0; i<8; i=i+1)
          begin
               {inputA, inputB, inputC} = i;
               for (j=0; j<8; j=j+1)
                    begin
                         inputD = j;
                         #10;
                    end
          end
end

bitwise bitwise_inst (.inputA(inputA), .inputB(inputB),
.inputC(inputC), .inputD(inputD), .outputA(outputA),
.outputB(outputB), .outputC(outputC),
.outputD(outputD));
```

```
initial
begin
    $monitor ("inputA %b inputB %b inputC %b inputD
%h outputA %b outputB %b outputC %b outputD
%h",inputA, inputB, inputC, inputD, outputA, outputB,
outputC, outputD);
end

endmodule
```

Example 4.45 Simulation Results for Verilog Test Bench Modulus "*bitwise_tb*"

```
inputA 0 inputB 0 inputC 0 inputD 0 outputA 0 outputB
0 outputC 1 outputD 0
inputA 0 inputB 0 inputC 0 inputD 1 outputA 0 outputB
0 outputC 1 outputD 0
inputA 0 inputB 0 inputC 0 inputD 2 outputA 0 outputB
0 outputC 1 outputD 0
inputA 0 inputB 0 inputC 0 inputD 3 outputA 0 outputB
0 outputC 1 outputD 0
inputA 0 inputB 0 inputC 0 inputD 4 outputA 0 outputB
0 outputC 1 outputD 0
inputA 0 inputB 0 inputC 0 inputD 5 outputA 0 outputB
0 outputC 1 outputD 0
inputA 0 inputB 0 inputC 0 inputD 6 outputA 0 outputB
0 outputC 1 outputD 0
inputA 0 inputB 0 inputC 0 inputD 7 outputA 0 outputB
0 outputC 1 outputD 0
inputA 0 inputB 0 inputC 1 inputD 0 outputA 0 outputB
0 outputC 0 outputD 0
inputA 0 inputB 0 inputC 1 inputD 1 outputA 0 outputB
0 outputC 0 outputD 1
inputA 0 inputB 0 inputC 1 inputD 2 outputA 0 outputB
0 outputC 0 outputD 0
inputA 0 inputB 0 inputC 1 inputD 3 outputA 0 outputB
0 outputC 0 outputD 1
inputA 0 inputB 0 inputC 1 inputD 4 outputA 0 outputB
0 outputC 0 outputD 0
inputA 0 inputB 0 inputC 1 inputD 5 outputA 0 outputB
0 outputC 0 outputD 1
inputA 0 inputB 0 inputC 1 inputD 6 outputA 0 outputB
0 outputC 0 outputD 0
inputA 0 inputB 0 inputC 1 inputD 7 outputA 0 outputB
0 outputC 0 outputD 1
```

```
inputA 0 inputB 1 inputC 0 inputD 0 outputA 0 outputB
1 outputC 1 outputD 0
inputA 0 inputB 1 inputC 0 inputD 1 outputA 0 outputB
1 outputC 1 outputD 0
inputA 0 inputB 1 inputC 0 inputD 2 outputA 0 outputB
1 outputC 1 outputD 2
inputA 0 inputB 1 inputC 0 inputD 3 outputA 0 outputB
1 outputC 1 outputD 2
inputA 0 inputB 1 inputC 0 inputD 4 outputA 0 outputB
1 outputC 1 outputD 0
inputA 0 inputB 1 inputC 0 inputD 5 outputA 0 outputB
1 outputC 1 outputD 0
inputA 0 inputB 1 inputC 0 inputD 6 outputA 0 outputB
1 outputC 1 outputD 2
inputA 0 inputB 1 inputC 0 inputD 7 outputA 0 outputB
1 outputC 1 outputD 2
inputA 0 inputB 1 inputC 1 inputD 0 outputA 0 outputB
1 outputC 0 outputD 0
inputA 0 inputB 1 inputC 1 inputD 1 outputA 0 outputB
1 outputC 0 outputD 1
inputA 0 inputB 1 inputC 1 inputD 2 outputA 0 outputB
1 outputC 0 outputD 2
inputA 0 inputB 1 inputC 1 inputD 3 outputA 0 outputB
1 outputC 0 outputD 3
inputA 0 inputB 1 inputC 1 inputD 4 outputA 0 outputB
1 outputC 0 outputD 0
inputA 0 inputB 1 inputC 1 inputD 5 outputA 0 outputB
1 outputC 0 outputD 1
inputA 0 inputB 1 inputC 1 inputD 6 outputA 0 outputB
1 outputC 0 outputD 2
inputA 0 inputB 1 inputC 1 inputD 7 outputA 0 outputB
1 outputC 0 outputD 3
inputA 1 inputB 0 inputC 0 inputD 0 outputA 0 outputB
1 outputC 1 outputD 0
inputA 1 inputB 0 inputC 0 inputD 1 outputA 0 outputB
1 outputC 1 outputD 0
inputA 1 inputB 0 inputC 0 inputD 2 outputA 0 outputB
1 outputC 1 outputD 0
inputA 1 inputB 0 inputC 0 inputD 3 outputA 0 outputB
1 outputC 1 outputD 0
inputA 1 inputB 0 inputC 0 inputD 4 outputA 0 outputB
1 outputC 1 outputD 4
inputA 1 inputB 0 inputC 0 inputD 5 outputA 0 outputB
1 outputC 1 outputD 4
```

```
inputA 1 inputB 0 inputC 0 inputD 6 outputA 0 outputB
1 outputC 1 outputD 4
inputA 1 inputB 0 inputC 0 inputD 7 outputA 0 outputB
1 outputC 1 outputD 4
inputA 1 inputB 0 inputC 1 inputD 0 outputA 0 outputB
1 outputC 0 outputD 0
inputA 1 inputB 0 inputC 1 inputD 1 outputA 0 outputB
1 outputC 0 outputD 1
inputA 1 inputB 0 inputC 1 inputD 2 outputA 0 outputB
1 outputC 0 outputD 0
inputA 1 inputB 0 inputC 1 inputD 3 outputA 0 outputB
1 outputC 0 outputD 1
inputA 1 inputB 0 inputC 1 inputD 4 outputA 0 outputB
1 outputC 0 outputD 4
inputA 1 inputB 0 inputC 1 inputD 5 outputA 0 outputB
1 outputC 0 outputD 5
inputA 1 inputB 0 inputC 1 inputD 6 outputA 0 outputB
1 outputC 0 outputD 4
inputA 1 inputB 0 inputC 1 inputD 7 outputA 0 outputB
1 outputC 0 outputD 5
inputA 1 inputB 1 inputC 0 inputD 0 outputA 1 outputB
1 outputC 1 outputD 0
inputA 1 inputB 1 inputC 0 inputD 1 outputA 1 outputB
1 outputC 1 outputD 0
inputA 1 inputB 1 inputC 0 inputD 2 outputA 1 outputB
1 outputC 1 outputD 2
inputA 1 inputB 1 inputC 0 inputD 3 outputA 1 outputB
1 outputC 1 outputD 2
inputA 1 inputB 1 inputC 0 inputD 4 outputA 1 outputB
1 outputC 1 outputD 4
inputA 1 inputB 1 inputC 0 inputD 5 outputA 1 outputB
1 outputC 1 outputD 4
inputA 1 inputB 1 inputC 0 inputD 6 outputA 1 outputB
1 outputC 1 outputD 6
inputA 1 inputB 1 inputC 0 inputD 7 outputA 1 outputB
1 outputC 1 outputD 6
inputA 1 inputB 1 inputC 1 inputD 0 outputA 1 outputB
1 outputC 0 outputD 0
inputA 1 inputB 1 inputC 1 inputD 1 outputA 1 outputB
1 outputC 0 outputD 1
inputA 1 inputB 1 inputC 1 inputD 2 outputA 1 outputB
1 outputC 0 outputD 2
inputA 1 inputB 1 inputC 1 inputD 3 outputA 1 outputB
1 outputC 0 outputD 3
```

```
inputA 1 inputB 1 inputC 1 inputD 4 outputA 1 outputB
1 outputC 0 outputD 4
inputA 1 inputB 1 inputC 1 inputD 5 outputA 1 outputB
1 outputC 0 outputD 5
inputA 1 inputB 1 inputC 1 inputD 6 outputA 1 outputB
1 outputC 0 outputD 6
inputA 1 inputB 1 inputC 1 inputD 7 outputA 1 outputB
1 outputC 0 outputD 7
```

The simulation results in Example 4.45 show that the bitwise operation performed on the operands is done bit-by-bit. Basically, the Verilog code for the vector format operation:

```
assign outputD = {inputA, inputB, inputC} & inputD;
```

is equivalent to:

```
assign outputD[2] = inputA & inputD [2];
assign outputD[1] = inputB & inputD [1];
assign outputD[0] = inputC & inputD [0];
```

4.8.9 Equality Operator

The equality operators are used for comparison of operands for equality. In Verilog, there are two types of equality operators:

1. **Logical equality** is represented by the symbols "==" for equal and "!=" for not equal. These symbols are often used in Verilog coding. The logical equality can produce results that are either 0 (false), 1 (true), or X (unknown). The result of X may occur when any of the operands used have either an X (unknown) or Z (tri-state). If an operand A has a four-bit value of "1001" and another operand B has a four-bit value of "1010," an operation of "A == B" would give a result of "0" (false). On the other hand, an operation of "A != B" would give a result of "1" (true).
2. **Case equality** is represented by the symbols "===" for equal and "!==" for not equal. The case equality always produces results that are either a logic 1 (true) or logic 0 (false). It cannot produce a result that is X (unknown). Case equality operators treat operands that have X or Z as values just as it treats 0 and 1. If an operand A has a four-bit value of "1xz0" and another operand B has a 4 bit value of "1xz0," an operation of "A === B" would give a result of "1" (true). On the other hand, an operation of "A !== B" would give a result of "0" (false). Designers need to take note that case equality operators are nonsynthesizable, as it would be impossible to create logic that has the functionality of detecting X (unknown) or Z (tristate) in the operands.

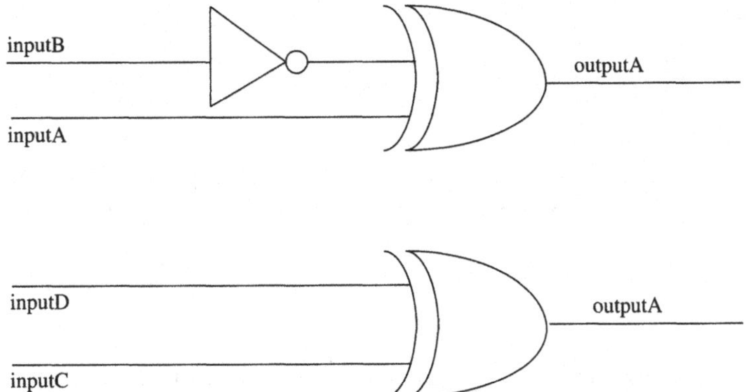

FIGURE 4.27. Diagram showing synthesized logic for verilog code module "***logicequal***."

Example 4.46 is a Verilog code that uses the logic equality operators. Figure 4.27 shows the synthesized logic for the Verilog code in Example 4.46.

Example 4.46 Verilog Code Using Logic Equality Operators

```
module logicequal (inputA, inputB, inputC, inputD,
outputA, outputB);

input inputA, inputB, inputC, inputD;
output outputA, outputB;

wire outputA, outputB;

assign outputA = (inputA == inputB);
assign outputB = (inputC != inputD);

endmodule
```

Example 4.47 shows a Verilog test bench that can be used to simulate the Verilog code module "***logicequal***" of Example 4.46 that uses logic equality operators. Example 4.48 shows the simulation results of the Verilog test bench module "***logicequal_tb***."

Example 4.47 Verilog Test Bench to Simulate Module "logicequal"

```
module logicequal_tb();

reg inputA, inputB, inputC, inputD;
wire outputA, outputB;
```

```
integer i;

initial
begin
     for (i=0; i<16; i=i+1)
          begin
               {inputA, inputB, inputC, inputD} = i;
               #10;
          end
end

logicequal logicequal_inst (.inputA(inputA),
.inputB(inputB),
.inputC(inputC), .inputD(inputD), .outputA(outputA),
.outputB(outputB));

initial
begin
     $monitor ("inputA %b inputB %b inputC %b inputD
%b outputA %b outputB %b", inputA, inputB, inputC,
inputD, outputA, outputB);
end

endmodule
```

Example 4.48 Simulation Results of Verilog Test Bench Module "*logicequal_tb*"

```
inputA 0 inputB 0 inputC 0 inputD 0 outputA 1 outputB 0
inputA 0 inputB 0 inputC 0 inputD 1 outputA 1 outputB 1
inputA 0 inputB 0 inputC 1 inputD 0 outputA 1 outputB 1
inputA 0 inputB 0 inputC 1 inputD 1 outputA 1 outputB 0
inputA 0 inputB 1 inputC 0 inputD 0 outputA 0 outputB 0
inputA 0 inputB 1 inputC 0 inputD 1 outputA 0 outputB 1
inputA 0 inputB 1 inputC 1 inputD 0 outputA 0 outputB 1
inputA 0 inputB 1 inputC 1 inputD 1 outputA 0 outputB 0
inputA 1 inputB 0 inputC 0 inputD 0 outputA 0 outputB 0
inputA 1 inputB 0 inputC 0 inputD 1 outputA 0 outputB 1
inputA 1 inputB 0 inputC 1 inputD 0 outputA 0 outputB 1
inputA 1 inputB 0 inputC 1 inputD 1 outputA 0 outputB 0
inputA 1 inputB 1 inputC 0 inputD 0 outputA 1 outputB 0
inputA 1 inputB 1 inputC 0 inputD 1 outputA 1 outputB 1
inputA 1 inputB 1 inputC 1 inputD 0 outputA 1 outputB 1
inputA 1 inputB 1 inputC 1 inputD 1 outputA 1 outputB 0
```

4.8.10 Reduction Operator

Reduction operators have the same functionality as that of logical operators except that reduction operators operate on the bits of the operand itself. The results obtained from reduction operators are single bit. The different types of reduction operators allowed in Verilog:

1. & – reduction AND operation
2. | – reduction OR operation
3. ∧ – reduction XOR operation
4. ~& – reduction NAND operation
5. ~| – reduction NOR operation
6. ~∧ – reduction XNOR operation

Example 4.49 is a Verilog code that uses reduction operators. Figure 4.28 shows the synthesized logic for the Verilog code in Example 4.49.

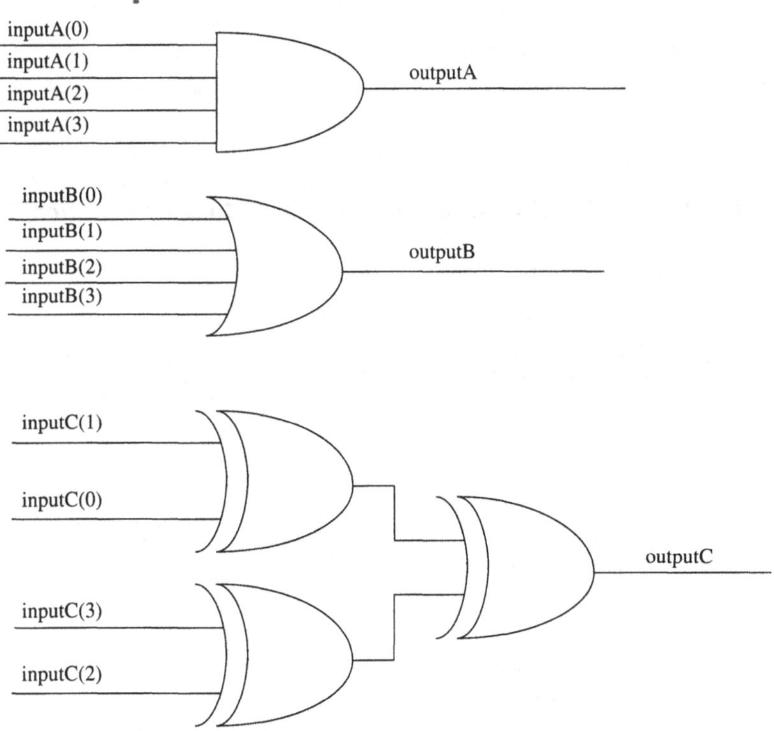

FIGURE 4.28. Diagram showing synthesized logic for verilog code module "***reduction.***"

Example 4.49 Verilog Code Using Reduction Operators

```
module reduction (inputA, inputB, inputC, outputA,
outputB, outputC);

input [3:0] inputA, inputB, inputC;
output outputA, outputB, outputC;

wire outputA, outputB, outputC;

// for reduction AND
assign outputA = &inputA;

// for reduction OR
assign outputB = |inputB;

// for reduction XOR
assign outputC = ^inputC;

endmodule
```

Example 4.50 shows a Verilog test bench that can be used to simulate the Verilog code module "***reduction***" of Example 4.49. Example 4.51 shows the simulation results.

Example 4.50 Verilog Test Bench for Simulating Module "*reduction*"

```
module reduction_tb ();

reg [3:0] inputA, inputB, inputC;
wire outputA, outputB, outputC;

integer i;

initial
begin
     for (i=0; i<16; i=i+1)
          begin
               inputA = i;
               inputB = i;
               inputC = i;
               #10;
          end
end
```

```
reduction reduction_inst (.inputA(inputA),
.inputB(inputB), .inputC(inputC), .outputA(outputA),
.outputB(outputB), .outputC(outputC));

initial
begin
    $monitor ("inputA %h inputB %h inputC %h outputA
%b outputB %b outputC %b", inputA, inputB, inputC,
outputA, outputB, outputC);
end

endmodule
```

Example 4.51. Simulation Results of Verilog Test Bench Module "*reduction_tb*"

```
inputA 0 inputB 0 inputC 0 outputA 0 outputB 0 outputC 0
inputA 1 inputB 1 inputC 1 outputA 0 outputB 1 outputC 1
inputA 2 inputB 2 inputC 2 outputA 0 outputB 1 outputC 1
inputA 3 inputB 3 inputC 3 outputA 0 outputB 1 outputC 0
inputA 4 inputB 4 inputC 4 outputA 0 outputB 1 outputC 1
inputA 5 inputB 5 inputC 5 outputA 0 outputB 1 outputC 0
inputA 6 inputB 6 inputC 6 outputA 0 outputB 1 outputC 0
inputA 7 inputB 7 inputC 7 outputA 0 outputB 1 outputC 1
inputA 8 inputB 8 inputC 8 outputA 0 outputB 1 outputC 1
inputA 9 inputB 9 inputC 9 outputA 0 outputB 1 outputC 0
inputA a inputB a inputC a outputA 0 outputB 1 outputC 0
inputA b inputB b inputC b outputA 0 outputB 1 outputC 1
inputA c inputB c inputC c outputA 0 outputB 1 outputC 0
inputA d inputB d inputC d outputA 0 outputB 1 outputC 1
inputA e inputB e inputC e outputA 0 outputB 1 outputC 1
inputA f inputB f inputC f outputA 1 outputB 1 outputC 0
```

4.8.11 Relational Operator

Relational operators are similar to equality operators except that relational operators return the compared result of relational value of equality. The result from using a relational operator is one bit. There are four types of relational operators:

1. **Greater than** This is represented by the symbol ">". It returns a result of value "1" (true) if an operand is conditionally greater in value than the other. If operand A has a value of 5, the condition "A > 3" would yield a result of 1, because the operand A is greater than 3.

2. **Less than** This is represented by the symbol "<". It returns a result of value "1" (true) if an operand is conditionally lesser in value than the other. If operand A has a value of 5, the condition "A < 3" would yield a result of 0, because the operand A is greater than 3.
3. **Greater than or equal** This is represented by the symbol ">=". It returns a result of value "1" (true) if an operand is conditionally greater or equal in value compared with the other. If operand A has a value of 3, the condition "A >= 3" would yield a result of 1, because the operand A is 3.
4. **Less than or equal** This is represented by the symbol "<=". It returns a result of value "1" (true) if an operand is conditionally lesser or equal in value compared with the other. If operand A has a value of 3, the condition "A <= 3" would yield a result of 1, because the operand A is 3.

Example 4.52 is a Verilog code that uses relational operators and Example 4.53 is a Verilog test bench that can be used for simulation.

Example 4.52 Verilog Code Using Relational Operators

```
module relational (inputA, inputB, inputC, inputD,
outputA, outputB, outputC, outputD, outputE, outputF,
outputG, outputH);

input [1:0] inputA, inputB, inputC, inputD;
output outputA, outputB, outputC, outputD, outputE,
outputF, outputG, outputH;

wire outputA, outputB, outputC, outputD, outputE,
outputF, outputG, outputH;

assign outputA = (inputA > 1);
assign outputB = (inputB < 2);
assign outputC = (inputC >= 1);
assign outputD = (inputD <= 2);

assign outputE = (inputA > inputB);
assign outputF = (inputB < inputC);
assign outputG = (inputC >= inputD);
assign outputH = (inputB <= inputD);

endmodule
```

Example 4.53 Verilog Test Bench for Simulation of Module "*relational*"

```
module relational_tb();
```

```
reg [1:0] inputA, inputB, inputC, inputD;
wire outputA, outputB, outputC, outputD, outputE,
outputF, outputG, outputH;

integer i,j;

initial
begin
     for (i=0; i<4; i=i+1)
     begin
          inputA = i;
          inputB = (3-i);
          for (j=0; j<4; j=j+1)
          begin
               inputC = j;
               inputD = (3-j);
               #10;
          end
     end
end

relational relational_inst (.inputA(inputA),
.inputB(inputB), .inputC(inputC), .inputD(inputD),
.outputA(outputA), .outputB(outputB),
.outputC(outputC), .outputD(outputD),
.outputE(outputE), .outputF(outputF),
.outputG(outputG), .outputH(outputH));

endmodule
```

Figure 4.29 shows the waveforms from the Verilog test bench simulation module "***relational_tb***."

Referring to the waveform in Figure 4.29:

1. ***assign outputA = (inputA > 1);*** ***outputA*** is at logic "1" (clock 9 to clock 16) when inputA has a value of either "2" or "3". This means that ***outputA*** is a logic "1" when the values at ***inputA*** is greater than "1", which is "2" or "3".
2. ***assign outputB = (inputB < 2);*** ***outputB*** is at logic "1" (clock 9 to clock 16) when ***inputB*** has a value of either "0" or "1". This means that ***outputB*** is a logic "1" when the values at ***inputB*** is less than "2", which is "1" or "0".
3. ***assign outputC = (inputC >= 1);*** ***outputC*** is at logic "1" when ***inputC*** has a value of either "1", "2", or "3". This means that ***outputC*** is a logic "1" when the values at ***inputC*** is greater than or equal to "1".

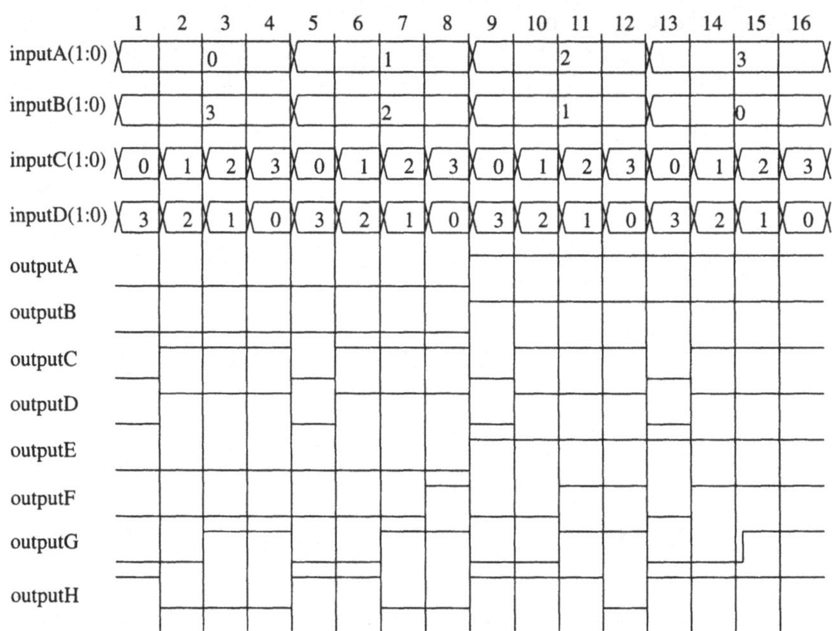

FIGURE 4.29. Diagram showing simulation results of Verilog test Bench module "*relational_tb*."

4. ***assign outputD = (inputD <= 2);*** ***outputD*** is at logic "1" when ***inputD*** has a value of either "0", "1", or "2". This means that ***outputD*** is a logic "1" when the values at inputD is less then or equal to "2".
5. ***assign outputE = (inputA > inputB);*** ***outputE*** is at logic "1" (clock 9 to clock 16) when:
   - ***inputA*** has a value of "2" and ***inputB*** has a value of "1" (clock 9 to clock 12)
   - ***inputA*** has a value of "3" and ***inputB*** has a value of "0" (clock 13 to clock 16)

   This means that ***outputE*** is a logic "1" when ***inputA*** is greater than ***inputB***.
6. ***assign outputF = (inputB < inputC);*** ***outputF*** is at logic "1" when:
   - ***inputB*** has a value of "2" and ***inputC*** has a value of "3" (clock 8)
   - ***inputB*** has a value of "1" and ***inputC*** has a value of "2" or "3" (clock 11, clock 12)
   - ***inputB*** has a value of "0" and ***inputC*** has a value of "1", "2", or "3" (clock 14, clock 15, clock 16)

   This means that ***outputF*** is a logic "1" when ***inputB*** is less than ***inputC***

7. ***assign outputG = (inputC >= inputD);*** ***outputG*** is at logic "1" when:
   - ***inputC*** has a value of "2" and ***inputD*** has a value of "1" (clock 3, clock 7, clock 11, clock 15)
   - ***inputC*** has a value of "3" and ***inputD*** has a value of "0" (clock 4, clock 8, clock 12, clock 16)

   This means that outputG is a logic "1" when ***inputC*** is greater than ***inputD***. However, the relational symbol used is ">=" (greater than or equal), but the Verilog test bench does not have a stimulus vector that tests for the condition of ***inputC*** = ***inputD***. This is a good example of an incomplete test bench because the test bench does not have stimulus to check for all conditions. It is important for the designer to note that, when writing Verilog test bench to check for a design, it is crucial that the designer tests for all the possible conditions that may occur.

8. ***assign outputH = (inputB <= inputD);*** ***outputH*** is at logic "1" when:
   - ***inputB*** has a value of "3" and ***inputD*** has a value of "3" (clock 1)
   - ***inputB*** has a value of "2" and ***inputD*** has a value of "3" (clock 5)
   - ***inputB*** has a value of "2" and ***inputD*** has a value of "2" (clock 6)
   - ***inputB*** has a value of "1" and ***inputD*** has a value of "3" (clock 9)
   - ***inputB*** has a value of "1" and ***inputD*** has a value of "2" (clock 10)
   - ***inputB*** has a value of "1" and ***inputD*** has a value of "1" (clock 11)
   - ***inputB*** has a value of "0" and ***inputD*** has a value of "3" (clock 13)
   - ***inputB*** has a value of "0" and ***inputD*** has a value of "2" (clock 14)
   - ***inputB*** has a value of "0" and ***inputD*** has a value of "1" (clock 15)
   - ***inputB*** has a value of "0" and ***inputD*** has a value of "0" (clock 16)

This means that ***outputH*** is a logic "1" when ***inputB*** is less than or equal to ***inputD***.

4.9 LATCH INFERENCE

When coding in Verilog, the designer needs to be careful to ensure that unwanted latches are not inferred. The condition of inferring unwanted latches often occurs when the designer uses "***if***" or "***case***" statements that are incomplete.

Example 4.54 is a Verilog code that uses an "***if***" statement to create combinational logic. However, because not all possible conditions are defined in the "***if***" statement, the value of ***outputA*** is maintained or latched for these undefined conditions. This is referred to as "inference of an unwanted latch."

Example 4.54 Verilog Code Using "if" Statement Inferring Unwanted Latch

```
module latch_infer (inputA, inputB, inputC, inputD,
outputA);

input inputA, inputB, inputC, inputD;
output outputA;

reg outputA;

always @ (inputA or inputB or inputC or inputD)
begin
     if (inputA & inputB)  ---->  Incomplete
          begin                   listing of
               if (inputC | ~inputD)  "if" conditions
                    outputA = 1'b1;   causing
               else                   inference of
                    outputA = 1'bZ;   latch.
          end
end

endmodule
```

Figure 4.30 shows a diagram for the synthesized logic module "***latch_infer***" in Example 4.54. The unwanted latch is inferred because the "***if***" statement that was used in Example 4.54, "***if (inputA & inputB)***," does not specify all other possible conditions. Therefore, the Verilog code executes the nested "***if***" statement when "***inputA***" and "***inputB***" are both at logical "1". But if either "***inputA***" or "***inputB***" is not at logical "1", there are no Verilog statements to direct the output signal ***outputA***. This means that the previous output value at ***outputA*** is maintained. In other words, a latch is inferred.

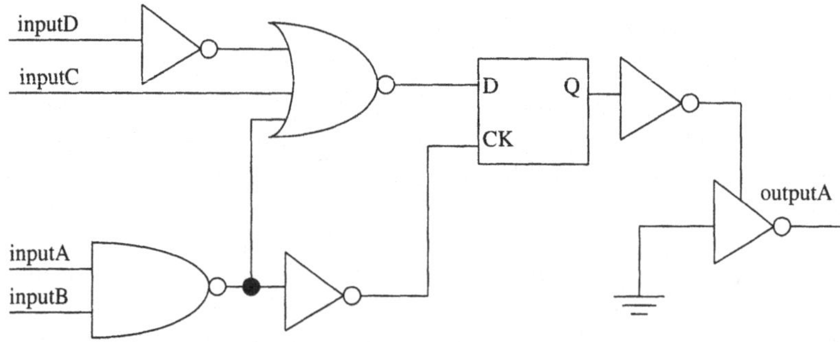

FIGURE 4.30. Diagram showing synthesized logic for module "*latch_infer*."

Example 4.55 shows a Verilog code that is the same as the code for module "***latch_infer***" in Example 4.54 except it has an "***else***" statement to direct the output value of ***outputA*** if either "***inputA***" or "***inputB***" is not at logical "1".

Example 4.55 Verilog Code Using "if" Statement That Does Not Infer Unwanted Latch

```
module latch_noninfer (inputA, inputB, inputC, inputD,
outputA);

input inputA, inputB, inputC, inputD;
output outputA;

reg outputA;

always @ (inputA or inputB or inputC or inputD)
begin
     if (inputA & inputB)
          begin
               if (inputC | ~inputD)
                    outputA = 1'b1;
               else
                    outputA = 1'bZ;
          end
     else
          outputA = 1'b0;
end

endmodule
```

"**else**" statement to specify the **outputA** signal if either **inputA** or **inputB** is not a logic "1".

The Verilog code for module "***latch_noninfer***" does not infer an unwanted latch, because all the possible conditions are specified. Therefore, the designer needs to remember that when coding in Verilog for synthesis, all possible conditions need to be specified. Figure 4.31 shows a diagram for the synthesized logic module "***latch_noninfer***" in Example 4.55.

Referring to Figures 4.30 and 4.31, notice how both the Verilog codes are synthesized to relatively different logic with only an additional "***else***" statement.

Apart from "***if***" statements, "***case***" statements may also cause conditions where unwanted latch is inferred. If a "***case***" statement is used, but does not declare all the possible conditions of the case statement, a latch will be inferred. Example 4.56 shows a Verilog code for module "***case_infer***," which uses the "***case***" statement but does not define all the possible conditions of the "***case***" statement.

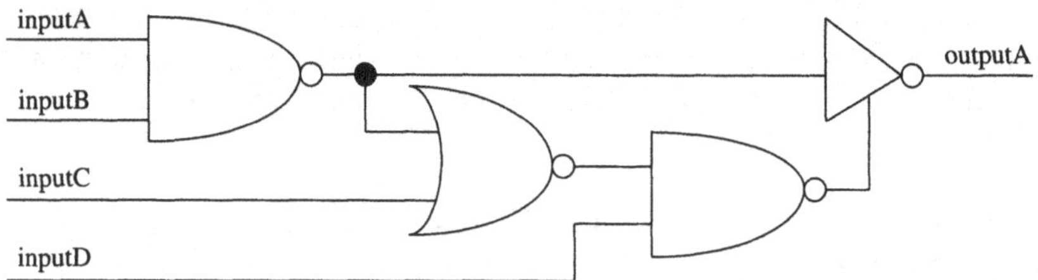

FIGURE 4.31. Diagram showing synthesized logic of module "*latch_noninfer*."

Example 4.56 Verilog Code for Incomplete "*case*" Statement

```
module case_infer (inputA, inputB, select, outputA);

input inputA, inputB ;
input [1:0] select;
output outputA;

reg outputA;

always @ (inputA or inputB or select)
begin
     case (select)
          2'b00: outputA = inputA;
          2'b01: outputA = inputB;
     endcase
end

endmodule
```

> Incomplete "**case**" condition, as only "00" and "01" conditions are defined. What happens when **select** is "10" or "11"?

Figure 4.32 shows a diagram for the synthesized logic module "***case_infer***." Notice how a latch is inferred at ***outputA***.

The inferred latch occurs in module "***case_infer***" because the Verilog code of the "***case***" statement does not specify the output signal value of ***outputA*** when ***select*** is a value other than "00" or "01". This would cause the previous value of ***outputA*** to be kept when ***select*** is neither "00" or "01".

A simple solution to the Verilog code of module "***case_infer***" would be to add a "***default***" condition in the "***case***" statement. The "***default***" condition would have ***outputA*** driven at logic "0" when ***select*** is neither "00" or "01".

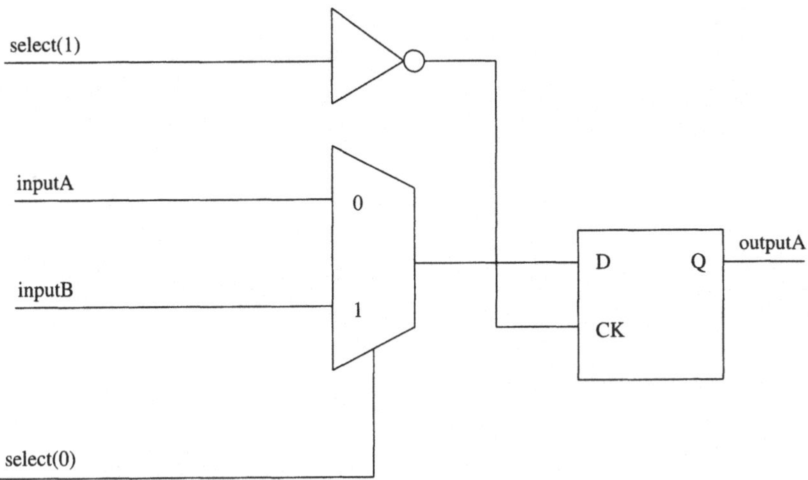

FIGURE 4.32. Diagram showing synthesized logic for module "***case_infer***."

Example 4.57 Verilog Code Utilizing "*default*" Condition for "*case*" Statement

```
module case_uninfer_diff (inputA, inputB, select,
outputA);

input inputA, inputB ;
input [1:0] select;
output outputA;

reg outputA;

always @ (inputA or inputB or select)
begin
     case (select)
          2'b00: outputA = inputA;
          2'b01: outputA = inputB;
          default: outputA = 1'b0;
     endcase
end

endmodule
```

"**default**" condition to define **outputA** when **select** is neither "00" or "01".

Figure 4.33 shows a diagram of the synthesized logic module "***case_uninfer_diff***." Notice that the synthesized logic no longer has a latch.

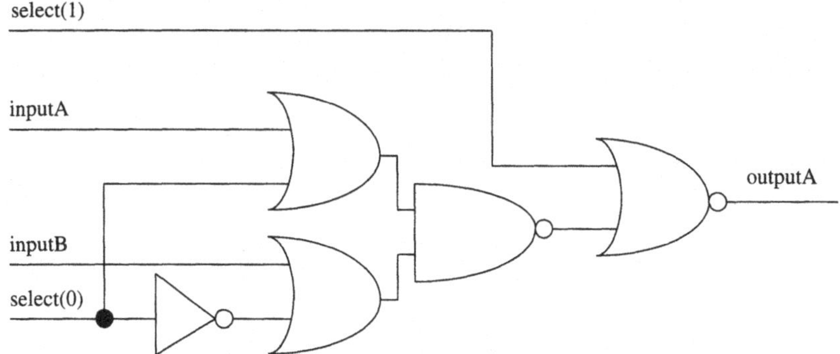

FIGURE 4.33. Diagram showing synthesized logic for module "***case_uninfer_diff***."

Apart from using the "***default***" condition as shown in the Verilog code of module "***case_uninfer_diff***," inference of a latch can also be avoided by specifing all the possible conditions of "***select***."

Example 4.58 Verilog Code Using "*case*" Statement with All Conditions Specified

```
module case_uninfer (inputA, inputB, select, outputA);

input inputA, inputB ;
input [1:0] select;
output outputA;

reg outputA;

always @ (inputA or inputB or select)
begin
     case (select)
          2'b00: outputA = inputA;
          2'b01: outputA = inputB;
          2'b10: outputA = 1'b0;
          2'b11: outputA = 1'b0;
     endcase
end
endmodule
```

All conditions of "**select**" is specified.

Module "***case_uninfer***" of Example 4.58 and module "***case_uninfer_diff***" of Example 4.57 both synthesizes to the same logic and has the same functionality.

The Verilog code shown in Examples 4.56, 4.57, and 4.58 uses the "***case***" statement to obtain multiplexer functionality within an "***always***" block. The same functionality can be obtained by using a conditional operator.

Example 4.59 Verilog Code to Obtain Multiplexer Functionality Using Conditional Operator

```
module case_uninfer_assign (inputA, inputB, select,
outputA);

input inputA, inputB ;
input [1:0] select;
output outputA;

wire outputA;

assign outputA = select[1] ? 1'b0 : select[0] ? inputB
: inputA;

endmodule
```

4.10 MEMORY ARRAY

When coding in Verilog for synthesis, sometimes a designer may want to code a memory array. Coding a memory array is common in behavioral coding and synthesizable coding, but for synthesis, the memory array may be limited to only a small array.

In synthesis, when a one-bit memory cell is coded, it is synthesized to a multiplexer and a flip-flop. Representation of a one-bit memory cell with a multiplexer and a flip-flop is definitely a waste of silicon real estate. However, even though a memory cell is rather large in terms of die area, it is still common for a designer to code a memory array provided that the array is small. A good example would be when a designer needs a small set of registers to store certain values. Or a designer may also code a memory array for synthesis when designing a microcontroller or microprocessor that may need a small register file.

Figure 4.34 shows a diagram of a synthesized logic for a one-bit memory cell. Synthesizing a large array of memory cells is a waste of silicon area. However, synthesizing memory cells is rather simple when coding in synthesizable Verilog compared with schematic capturing of the same array.

Example 4.60 shows a Verilog code for a 1-kilobyte memory module that is synthesizable. However, the designer can take note that, although the Verilog code is simple and easy, synthesis of the code may take several minutes

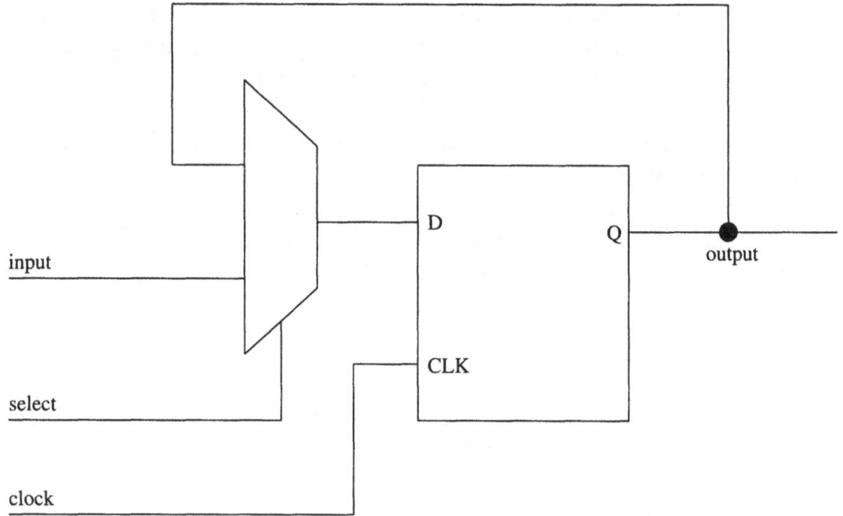

FIGURE 4.34. Diagram showing synthesis representation of a one-bit memory cell.

longer than it does on other simple Verilog code, because this short piece of code synthesizes to 1 kilobyte of memory.

Example 4.60 Verilog Code for a 1-kilobyte Memory Array

```
module memory (addr, data_in, data_out, write, read,
clock, reset);

// 1kbyte memory module - 128 address x 8 bits

input [6:0] addr;
input [7:0] data_in;
input write, read, clock, reset;
output [7:0] data_out;

reg [7:0] data_out;
reg [7:0] memory [127:0];

integer i;

// asynchronous reset
always @ (posedge clock or posedge reset)
begin
     if (reset)
          begin
```

```
                data_out = 0;
                // to initialize all memory to zero
                for (i=0; i<128; i=i+1)
                     memory[i] <= 0;
          end
     else
          begin
                if (read)
                     data_out <= memory [addr];
                else if (write)
                     begin
                          data_out <= 0;
                          memory [addr] <= data_in;
                     end
          end
end

endmodule
```

Example 4.61 shows a Verilog test bench that can be used to simulate the memory module shown in Example 4.60.

Example 4.61 Verilog Test Bench Module "*memory_tb*" to Simulate Module "*memory*"

```
module memory_tb ();

reg [6:0] address, addr;
reg [7:0] data, data_in;
reg write, read, clock, reset;
wire [7:0] data_out;

parameter cycle = 20;

integer i;

initial
begin
     addr = 0;
     reset = 0;
     read = 0;
     write = 0;
     data_in = 0;
     clock = 0;
     forever #20 clock = ~clock;
end
```

```
initial
begin
     // for reset
     reset = 0;
     #cycle;
     reset = 1;
     #cycle;
     reset = 50;
     #cycle;

     for (i=10; i<15; i=i+1)
          begin
               address = i;
               data = i;
               memory_write (address, data);
               #cycle;
          end

     for (i=14; i>=10; i=i-1)
          begin
               address = i;
               memory_read (address);
               #cycle;
          end
     $stop;
end

task memory_write;
input [7:0] data;
input [6:0] address;
begin
     addr = address;
     data_in = data;
     write = 0;
     #cycle;
     write = 1;
     read = 0;
     repeat (2) #cycle;
     write = 0;
     #cycle;
     $display ("Completed writing data %h at address
%h", data_in, addr);
end
endtask
```

Declaration of task "**memory_write**" for writing data to memory at location "**address**"

```
task memory_read;
input [6:0] address;
begin
     read = 0;
     addr = address;
     data_in = 0;
     #cycle;
     read = 1;
     repeat (2) #cycle;
     read = 0;
     #cycle;
     $display ("Completed reading memory at address
%h. Data is %h", addr, data_out);
end
endtask

memory memory_inst (addr, data_in, data_out, write,
read, clock, reset);

endmodule
```

Task "***memory_read***" for reading memory from location "***address***"

Figure 4.35 is a diagram showing the waveform results of Verilog test bench module "***memory_tb***."

Referring to Figure 4.35:

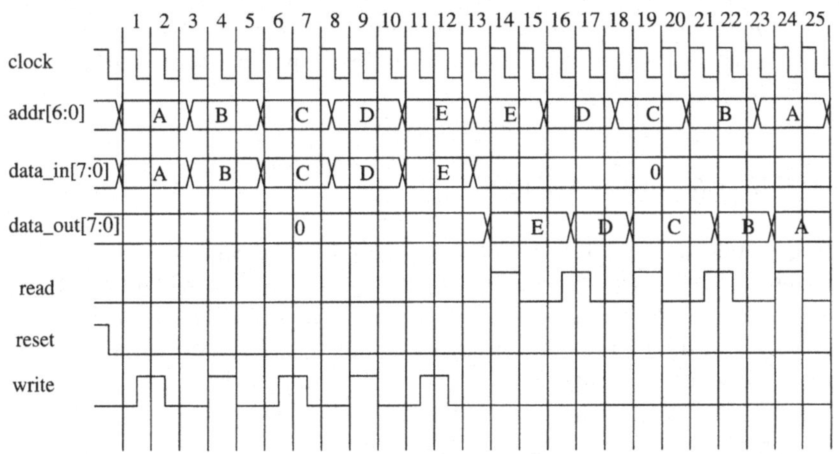

FIGURE 4.35. Diagram showing simulation waveform for Verilog test bench module "*memory_tb*"

1. On rising edge of ***clock*** 1, 3, 5, 6, 8, 10, 11, and 13, ***read*** and ***write*** is at logical "0". No reading or writing to memory occurs.
2. On rising edge of ***clock*** 2, 4, 7, 9, and 12, ***write*** is at a logical "1". A memory write occurs at address ***addr(6:0)*** with data ***data_in(7:0)***.
   - On rising edge of ***clock*** 2, ***addr(6:0)*** = "A", ***data_in(7:0)*** = "A", ***write*** = '1' → memory write at address "A" with data "A".
   - On rising edge of ***clock*** 4, ***addr(6:0)*** = "B", ***data_in(7:0)*** = "B", ***write*** = "1" → memory write at address "B" with data "B".
   - On rising edge of ***clock*** 7, ***addr(6:0)*** = "C", ***data_in(7:0)*** = "C", ***write*** = "1" → memory write at address "C" with data "C".
   - On rising edge of ***clock*** 9, ***addr(6:0)*** = "D", ***data_in(7:0)*** = "D", ***write*** = "1" → memory write at address "D" with data "D".
   - On rising edge of ***clock*** 12, ***addr(6:0)*** = "E", ***data_in(7:0)*** = "E", ***write*** = "1" → memory write at address "E" with data "E".
3. On rising edge of ***clock*** 15, 16, 18, 20, 21, 23, and 25 ***read*** and ***write*** is at logical "0". No reading or writing to memory occurs.
4. On rising edge of ***clock*** 14, 17, 19, 22, and 24, ***read*** is at a logical "1". A memory read occurs at address ***addr(6:0)*** and output ***data_out(7:0)*** is driven with data from the memory module.
   - On rising edge of ***clock*** 14, ***addr(6:0)*** = "E", ***read*** = "1" → memory read at address "E". Data read is "E" and is shown on ***data_out(7:0)***.
   - On rising edge of ***clock*** 17, ***addr(6:0)*** = "D", ***read*** = "1" → memory read at address "D". Data read is "D" and is shown on ***data_out(7:0)***.
   - On rising edge of ***clock*** 19, ***addr(6:0)*** = "C", ***read*** = "1" → memory read at address "C". Data read is "C" and is shown on ***data_out(7:0)***.
   - On rising edge of ***clock*** 22, ***addr(6:0)*** = "B", ***read*** = "1" → memory read at address "B". Data read is "B" and is shown on ***data_out(7:0)***.
   - On rising edge of ***clock*** 24, ***addr(6:0)*** = "A", ***read*** = "1" → memory read at address "A". Data read is "A" and is shown on ***data_out(7:0)***.

4.11 STATE MACHINE DESIGN

A frequent design found in almost all types of digital design is a state machine. It is easy to design and gives the designer great flexibility when the designer needs to tweak the design either for speed or area optimization. Most synthesis tools in the market have special options to allow a designer to synthesize a state machine design. These options allow the designer to easily choose the form of state machine implementation, whether a binary encoding, gray encoding, or one hot encoding.

A state machine is always coded using a "***case***" statement. This statement allows for multiple conditions to be made for different decisions.

Designs that normally consist of a state machine are designs that have a fixed amount of known states, meaning that there is a fixed amount of states, with each state having its own functionality to perform.

Note: Most synthesis tools, such as Synopsys's Design Compiler, have special built-in algorithms for synthesizing state machine design (Synopsys's Design Compiler has a built-in Finite State Machine Compiler that is specially used for synthesizing and tweaking state machine). These special algorithm allow the designer to choose different types of encoding as well as different types of optimization to obtain the most optimal synthesis result, either in the form of area or performance. To have a better understanding of how Synopsys's Design Compiler (or rather Finite State Machine Compiler) can be used for optimizing state machine-based designs, please refer to VHDL Coding and Logic Synthesis with Synopsys, by ***Weng Fook Lee****, published by* ***Academic Press****.*

4.11.1 Intelligent Traffic Control System

A good example of a design that can be coded in state machine format is that of an intelligent traffic control system. A traffic control system has only certain fixed conditions to fulfill in order to control the traffic lights that control traffic flow.

Figure 4.36 shows an interchange junction of four roads. Traffic crossing the interchange junction needs to be regulated by a traffic light system in order to enable a smooth and safe interchange crossing for motorized vehicles. Now, assume that a design is needed to control the traffic light system.

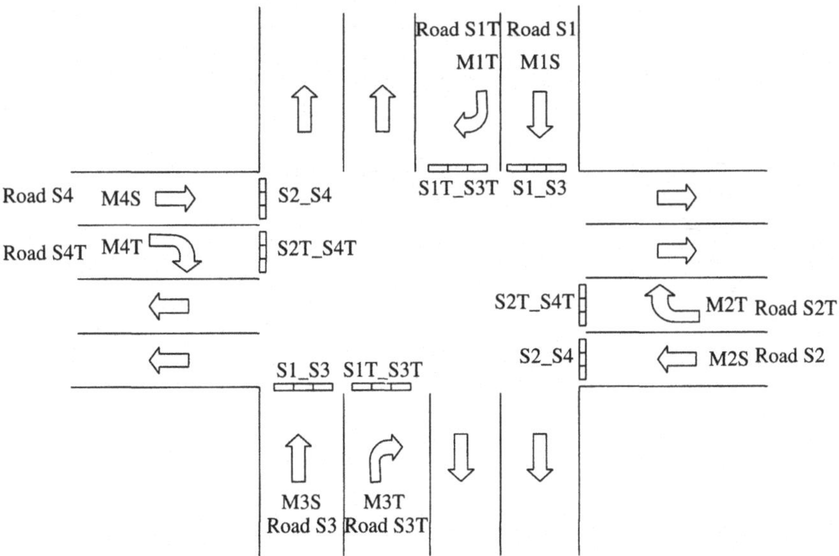

FIGURE 4.36. Diagram showing an interchange junction traffic light.

Referring to Figure 4.36, the interchange junction has four sets of traffic lights, ***S1_S3***, ***S2_S4***, ***S1T_S3T***, and ***S2T_S4T***. These four sets of traffic lights allow four different passes on the interchange junction.

Figure 4.37 shows a possible scenario whereby the traffic lights of set ***S1_S3*** are ***GREEN*** while all others are ***RED***.

Figure 4.38 shows a possible second scenario whereby the traffic lights of set ***S2_S4*** are ***GREEN*** while all others are ***RED***.

Figure 4.39 shows a possible third scenario whereby the traffic lights of set ***S1T_S3T*** are ***GREEN*** while all others are ***RED***.

Figure 4.40 shows a possible fourth scenario whereby the traffic lights of set ***S2T_S4T*** are ***GREEN*** while all others are ***RED***.

Figures 4.37 through 4.40 show the possible passes through the interchange junction. Apart from the four sets of traffic lights of ***S1_S3***, ***S2_S4***, ***S1T_S3T***, and ***S2T_S4T***, there are present eight sensors, ***M1S***, ***M1T***, ***M2S***, ***M2T***, ***M3S***, ***M3T***, ***M4S***, and ***M4T***. Each sensor is strategically located in each road to sense whether there are any cars in queue. For example, if there are cars queuing to turn from road ***S2T*** or road ***S4T*** (as shown in Fig. 4.40), the sensor ***M2T*** would sense that cars are queuing at road ***S2T*** (or cars are queuing at road ***S4T***, respectively). Similarly, the same for the other sensors. Sensors ***M1S*** would sense for cars queuing at road ***S1***, ***M1T*** would sense for cars queuing at road ***S1T***, sensor ***M2S*** would sense for cars queuing at road ***S2*** and so forth. These eight sensors together with an external timer would allow for the building block of an "intelligent" traffic light system.

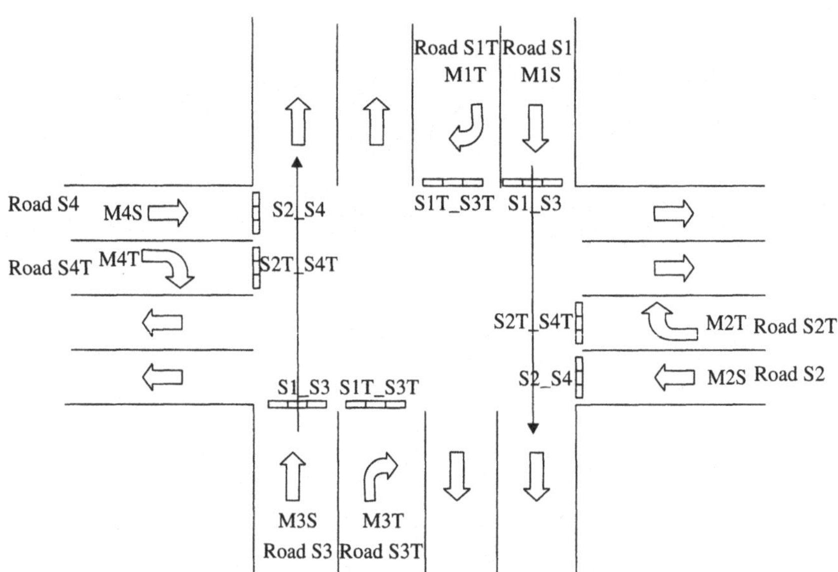

FIGURE 4.37. Diagram showing possible pass scenario whereby *S1_S3* is GREEN.

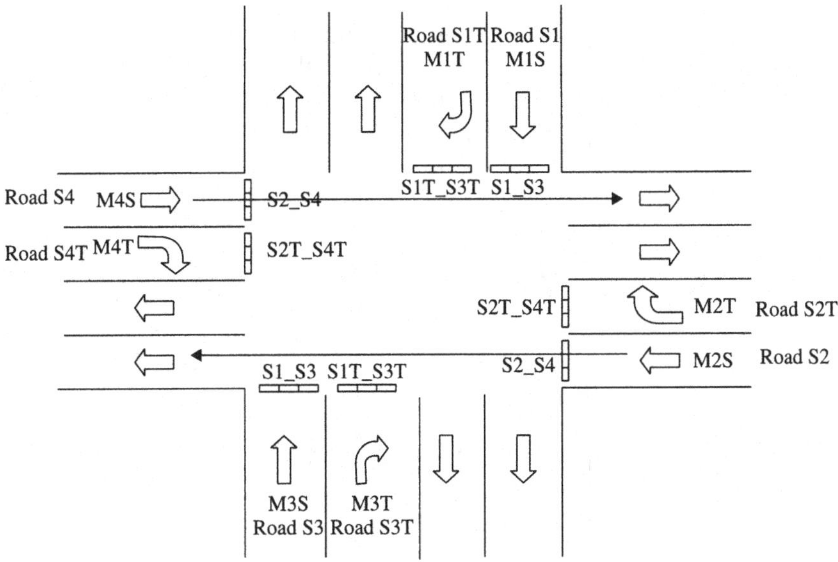

FIGURE 4.38. Diagram showing possible pass scenario whereby *S2_S4* is GREEN.

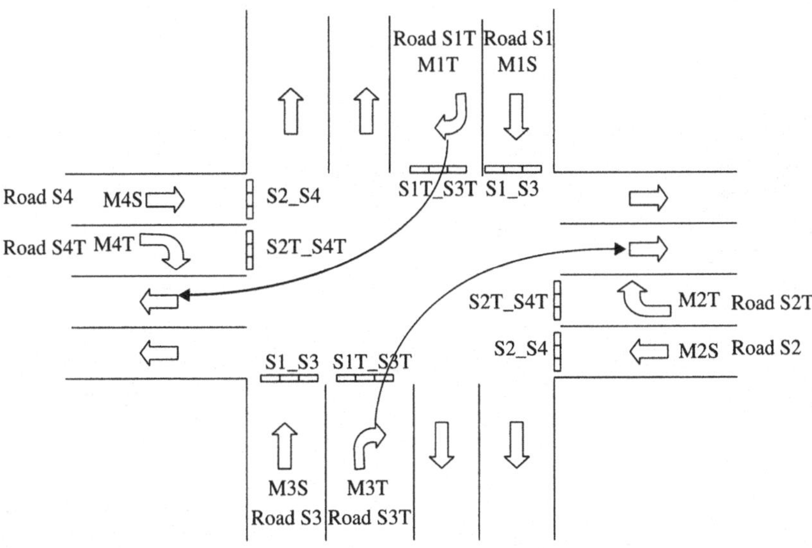

FIGURE 4.39. Diagram showing possible pass scenario whereby *S1T_S3T* is GREEN.

Based on the requirements stated, an interface block is created that defines the input and output signals of the traffic system controller. Figure 4.41 shows a figure of the input and outputs signals for the traffic light controller. Table 4.3 shows a description of the functions of each interface signal.

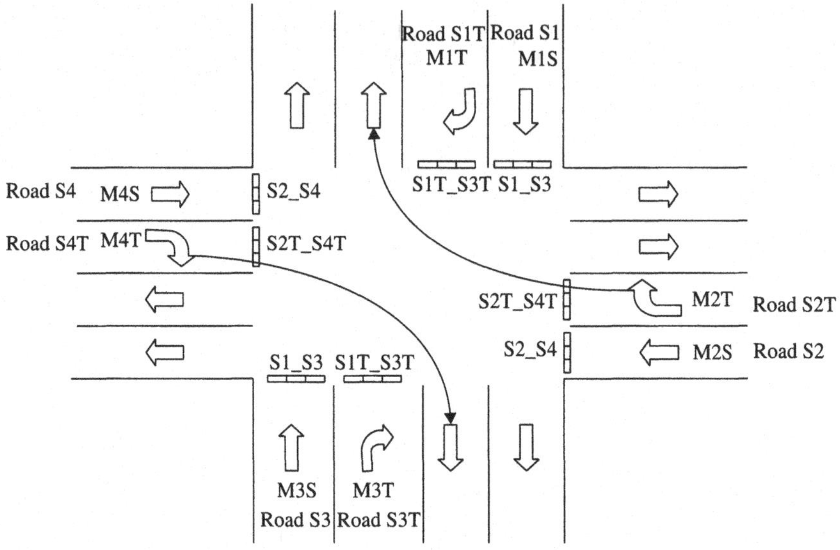

FIGURE 4.40. Diagram showing possible pass scenario whereby *S2T_S4T* is GREEN.

clock
reset
M1S
M2S
M3S
M4S
M1T
M2T
M3T
M4T
timerGreen
timerYellow
timerRed

TRAFFIC
LIGHT
STATE
MACHINE
CONTROLLER

S1_S3
S2_S4
S1T_S3T
S2T_S4T

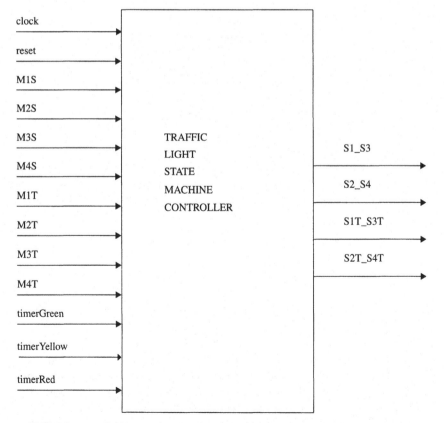

FIGURE 4.41. Diagram showing interface signals for traffic light controller.

TABLE 4.3. Interface Signal Description for Traffic Light Controller

Signal Name	Direction	Description
clock	Input	Clock input to the state machine
reset	Input	Reset signal to "reset" the state machine to a known state
M1S	Input	Sensor to sense for cars queuing in road ***S1***
M2S	Input	Sensor to sense for cars queuing in road ***S2***
M3S	Input	Sensor to sense for cars queuing in road ***S3***
M4S	Input	Sensor to sense for cars queuing in road ***S4***
M1T	Input	Sensor to sense for cars queuing in road ***S1T***
M2T	Input	Sensor to sense for cars queuing in road ***S2T***
M3T	Input	Sensor to sense for cars queuing in road ***S3T***
M4T	Input	Sensor to sense for cars queuing in road ***S4T***
timerGreen	Input	Input signal from a timer that times out after certain period to "inform" the state machine to change from ***GREEN*** to ***YELLOW***
timerYellow	Input	Input signal from a timer that times out after certain period to "inform" the state machine to change from ***YELLOW*** to ***RED***
timerRed	Input	Input signal from a timer that times out after certain period to "inform" the state machine to change from ***RED*** to ***GREEN*** (assuming there is traffic)
S1_S3	Output	Two-bit bus that controls the traffic lights for road ***S1*** and ***S3***. "01"—***GREEN*** "10"—***YELLOW*** "11"—***RED***
S2_S4	Output	Two-bit bus that controls the traffic lights for road ***S2*** and ***S4***. "01"—***GREEN*** "10"—***YELLOW*** "11"—***RED***
S1T_S3T	Output	Two-bit bus that controls the traffic lights for road ***S1T*** and ***S3T***. "01"—***GREEN*** "10"—***YELLOW*** "11"—***RED***
S2T_S4T	Output	Two-bit bus that controls the traffic lights for road ***S2T*** and ***S4T***. "01"—***GREEN*** "10"—***YELLOW*** "11"—***RED***

*Note: The inputs **timerGreen, timerYellow,** and **timerRed** are assumed to be generated from an external timer module that allows the traffic light controller to "know" that the allocated time period for the traffic lights at **GREEN, YELLOW,** or **RED** is exhausted and the traffic lights are to switch colors. It is also assumed that this timer module takes the inputs from the eight sensors to allow it to automatically determine if a certain traffic light should be at **GREEN, YELLOW,** or **RED** for a longer or shorter period of time before switching to another color.*

Based on the interface signals as shown in Table 4.3 as well as the different conditions shown in Figures 4.37 through 4.40, a state diagram is drawn to reflect all the possible conditions or situations that may happen. The traffic light state machine controller must be able to handle all of the possible conditions. Figure 4.42 shows the state diagram for the traffic light state machine controller.

From Figure 4.42, each of the state transitions are represented by the following conditions:

```
A = timerRed=1 & (M3S=1 | M1S=1)
B = timerRed=1 & M1S=0 & M3S=0 & (M1T=1 | M3T=1)
C = timerRed=1 & M1S=0 & M3S=0 & M1T=0 & M3T=0 &
(M2S=1 | M4S=1)
D = timerRed=1 & M1S=0 & M3S=0 & M1T=0 & M3T=0 &
M2S=0 & M4S=0 & (M2T=1 | M4T=1)
```

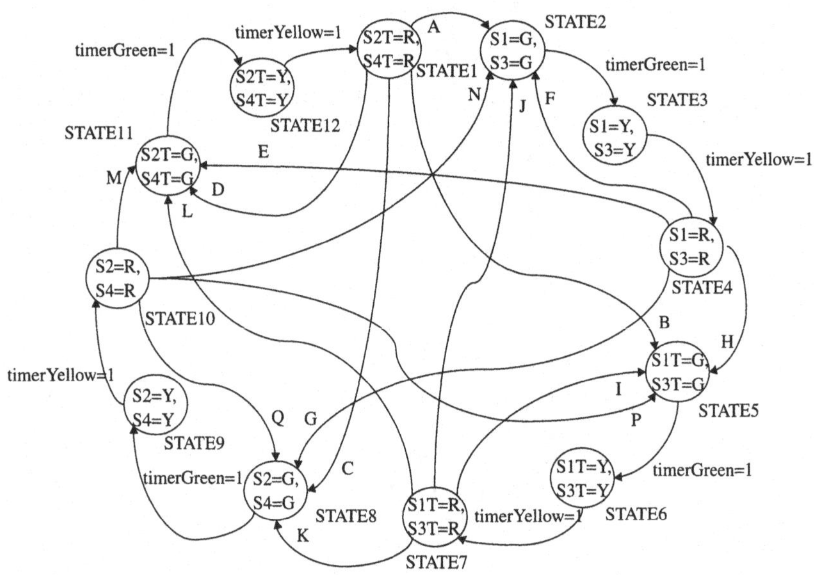

FIGURE 4.42. State diagram for traffic light state machine controller.

```
E = timerRed=1 & (M1T=0 & M3T=0) &(M2S=0 & M4S=0) &
(M2T=1 | M4T=1)
F = timerRed=1 & (M1T=0 & M3T=0 & M2S=0 & M4S=0 &
M2T=0 & M4T=0 & (M1S=1 | M3S=1))
G = timerRed=1 & (M1T=0 & M3T=0) & (M2S=1 |  M4S=1)
H = timerRed=1 & (M1T=1 | M3T=1)
I = timerRed=1 & M2S=0 & M4S=0 & M2T=0 & M4T=0 &
M1S=0 & M3S=0 & (M1T=1 | M3T=1)
J = timerRed=1 & M2S=0 & M4S=0 & M2T=0 & M4T=0 &
(M1S=1 | M3S=1)
K = timerRed=1 & (M2S=1 | M4S=1)
L = timerRed=1 & M2S=0 & M4S=0 & (M2T=1 | M4T=1)
M = timerRed=1 & (M2T=1 | M4T=1)
N = timerRed=1 & M2T=0 & M4T=0 & (M1S=1 | M3S=1)
P = timerRed=1 & M2T=0 & M4T=0 & M1S=0 & M3S=0 &
(M1T=1 | M3T=1)
Q = timerRed=1 & M2T=0 & M4T=0 & M1S=0 & M3S=0 &
M1T=0 & M3T=0 & (M2S=1 | M4S=1)
```

Referring to Figure 4.42:

1. ***STATE1***, ***STATE4***, ***STATE7***, ***STATE10*** – all traffic lights are ***RED***.
2. In ***STATE1***, there are four possible state transitions:
   - State transition A occurs when ***timerRed*** is at logic “1” AND either sensor ***M1S*** OR ***M3S*** is at logic “1” ***(timerRed=1 & (M3S=1 | M1S=1))***. This means that state transition A occurs when the allocated period for the traffic lights at ***RED*** is exhausted and there are cars queuing either at road ***S1*** or ***S3***.
   - State transition B occurs when ***timerRed*** is at logic “1” AND sensor ***M1S***, ***M2S*** is at logical “0” AND either sensor ***M1T*** OR ***M3T*** is at logic “1” ***(timerRed=1 & M1S=0 & M3S=0 & (M1T=1 | M3T=1)***. This means that state transition B occurs when the allocated period for the traffic lights at ***RED*** is exhausted and there are cars queuing either at road ***S1T*** or ***S3T*** and there are no cars queuing at road ***S1*** and ***S3***.
   - State transition C occurs when ***timerRed*** is at logic “1” AND sensor ***M1S***, ***M3S***, ***M1T***, ***M3T*** is at logical “0” AND either sensor ***M2S*** or ***M4S*** at logic “1” ***(timerRed=1 & M1S=0 & M3S=0 & M1T=0 & M3T=0 & (M2S=1 | M4S=1))***. This means that state transition C occurs when the allocated period for the traffic lights at ***RED*** is exhausted and there are cars queuing either at road ***S2*** or ***S4*** and there are no cars queuing at road ***S1***, ***S3***, ***S1T***, and ***S3T***.
   - State transition D occurs when ***timerRed*** is at logic “1” AND sensors ***M1S***, ***M3S***, ***M1T***, ***M3T***, ***M2S***, and ***M4S*** are at logic “0” AND either

sensor ***M2T*** OR ***M4T*** is at logic "1" ***(timerRed=1 & M1S=0 & M3S=0 & M1T=0 & M3T=0 & M2S=0 & M4S=0 & (M2T=1 | M4T=1))***. This means that state transition D occurs when the allocated period for the traffic lights at ***RED*** is exhausted and there are cars queuing either at road ***S2T*** or ***S4T*** and there are no cars queuing at road ***S1***, ***S3***, ***S1T***, ***S3T***, ***S2***, and ***S4***.

3. In ***STATE2***, the traffic lights on road ***S1*** and road ***S3*** are ***GREEN***. ***STATE2*** will transition to ***STATE3*** when ***timerGreen*** is at logical "1" (the allocated period for the traffic lights at ***GREEN*** is exhausted).
4. In ***STATE3***, the traffic lights on road ***S1*** and road ***S3*** is ***YELLOW***. ***STATE3*** will transition to ***STATE4*** when ***timerYellow*** is at logical "1" (the allocated period for the traffic lights at ***YELLOW*** is exhausted).
5. In ***STATE4***, there are four possible state transitions.
   - State transition E occurs when ***timerRed*** is at logic "1" AND sensor ***M1T***, ***M3T***, ***M2S***, ***M4S*** is at logic "0" AND either sensor ***M2T*** OR ***M4T*** is at logic "1" ***(timerRed=1 & (M1T=0 & M3T=0) & (M2S=0 & M4S=0) & (M2T=1 | M4T=1))***. This means that state transition E occurs when the allocated period for the traffic lights at ***RED*** is exhausted and there are cars queuing either at road ***S2T*** or road ***S4T*** and there are no cars queuing at road ***S1T***, ***S3T***, ***S2***, and ***S4***.
   - State transition F occurs when ***timerRed*** is at logic "1" AND sensor ***M1T***, ***M3T***, ***M2S***, ***M4S***, ***M2T***, and ***M4T*** is at logic "0" AND either sensor ***M1S*** OR ***M3S*** is at logic "1" ***(timerRed=1 & (M1T=0 & M3T=0 & M2S=0 & M4S=0 & M2T=0 & M4T=0 & (M1S=1 | M3S=1))***. This means that state transition F occurs when the allocated period for the traffic lights at ***RED*** is exhausted and there are cars queuing either at road ***S1*** or road ***S3*** and there are no cars queuing at road ***S1T***, ***S3T***, ***S2***, ***S4***, ***S2T***, and ***S4T***.
   - State transition G occurs when ***timerRed*** is at logic "1" AND sensor ***M1T*** and ***M3T*** is at logic "0" AND either sensor ***M2S*** OR ***M4S*** is at logic "1" ***(timerRed=1 & (M1T=0 & M3T=0) & (M2S=1 | M4S=1))***. This means that state transition G occurs when the allocated period for the traffic lights at ***RED*** is exhausted and there are cars queuing either at road ***S2*** or road ***S4*** and there are no cars queuing at road ***S1T*** and ***S3T***.
   - State transition H occurs when ***timerRed*** is at logic "1" AND either sensor ***M1T*** OR ***M3T*** is at logic "1" ***(timerRed=1 & (M1T=1 | M3T=1))***. This means that state transition H occurs when the allocated period for the traffic lights at ***RED*** is exhausted and there are cars queuing at road ***S1T*** or road ***S3T***.
6. In STATE5, the traffic lights on road ***S1T*** and road ***S3T*** are ***GREEN***. ***STATE5*** will transition to ***STATE6*** when timerGreen is at logical "1" (the allocated period for the traffic lights at ***GREEN*** is exhausted).

7. In ***STATE6***, the traffic lights on road ***S1T*** and road ***S3T*** is ***YELLOW.*** ***STATE6*** will transition to ***STATE7*** when ***timerYellow*** is at logical "1" (the allocated period for the traffic lights at ***YELLOW*** is exhausted).
8. In ***STATE7***, there are four possible state transitions.
   - State transition I occurs when ***timerRed*** is at logic "1" AND sensor ***M2S***, ***M4S***, ***M2T***, ***M4T***, ***M1S***, and ***M3S*** is at logic "0" AND either sensor ***M1T*** OR ***M3T*** is at logic "1" ***(timerRed=1 & M2S=0 & M4S=0 & M2T=0 & M4T=0 & M1S=0 & M3S=0 & (M1T=1 | M3T=1))***. This means that state transition I occurs when the allocated period for the traffic lights at ***RED*** is exhausted and there are cars queuing at road ***S1T*** or road ***S3T*** and there are no cars queuing at road ***S2***, ***S4***, ***S2T***, ***S4T***, ***S1***, and ***S3***.
   - State transition J occurs when ***timerRed*** is at logic "1" AND sensor ***M2S, M4S, M2T and M4T*** is at logic "0" AND either sensor ***M1S*** OR ***M2S*** is at logic "1" ***(timerRed=1 & M2S=0 & M4S=0 & M2T=0 & M4T=0 & (M1S=1 | M3S=1))***. This means that state transition J occurs when the allocated period for the traffic lights at ***RED*** is exhausted and there are cars queuing at road ***S1*** or road ***S3*** and there are no cars queuing at road ***S2***, ***S4***, ***S2T***, and ***S4T***.
   - State transition K occurs when ***timerRed*** is at logic "1" AND either sensor ***M2S*** OR ***M4S*** is at logic "1" ***(timerRed=1 & (M2S=1 | M4S=1))***. This means that state transition K occurs when the allocated period for the traffic lights at ***RED*** is exhausted and there are cars queuing at road ***S2*** or road ***S4***.
   - State transition L occurs when ***timerRed*** is at logic "1" and sensor ***M2S*** and ***M4S*** is at logic "0" AND either sensor ***M2T*** OR ***M4T*** is at logic "1" ***(timerRed=1 & M2S=0 & M4S=0 & (M2T=1 | M4T=1))***. This means that state transition L occurs when the allocated period for the traffic lights at ***RED*** is exhausted and there are cars queuing at road ***S2T*** or road ***S4T*** and there are no cars queuing at road ***S2*** and ***S4***.
9. In ***STATE8***, the traffic lights on road ***S2*** and road ***S4*** are ***GREEN.*** ***STATE8*** will transition to ***STATE9*** when ***timerGreen*** is at logical "1" (the allocated period for the traffic lights at ***GREEN*** is exhausted).
10. In ***STATE9***, the traffic lights on road ***S2*** and road ***S4*** is ***YELLOW.*** ***STATE9*** will transition to ***STATE10*** when ***timerYellow*** is at logical "1" (the allocated period for the traffic lights at ***YELLOW*** is exhausted).
11. In ***STATE10***, there are four possible state transitions:
    - State transition M occurs when ***timerRed*** is at logic "1" AND either sensor ***M2T*** OR ***M4T*** is at logic "1" ***(timerRed=1 & (M2T=1 | M4T=1))***. This means that state transition M occurs when the allocated period for the traffic lights at ***RED*** is exhausted and there are cars queuing at road ***S2T*** or road ***S4T***.

- State transition N occurs when ***timerRed*** is at logic "1" AND sensor ***M2T*** and ***M4T*** is at logic "0" AND either sensor ***M1S*** OR ***M3S*** is at logic "1" ***(timerRed=1 & M2T=0 & M4T=0 & (M1S=1 | M3S=1))***. This means that state transition N occurs when the allocated period for the traffic lights at ***RED*** is exhausted and there are cars queuing at road ***S1*** or road ***S3*** and there are no cars queuing at road ***S2T*** and road ***S4T***.
- State transition P occurs when ***timerRed*** is at logic "1" AND sensor ***M2T***, ***M4T***, ***M1S***, and ***M3S*** are at logic "0" AND either sensor ***M1T*** OR ***M3T*** is at logic "1" ***(timerRed=1 & M2T=0 & M4T=0 & M1S=0 & M3S=0 & (M1T=1 | M3T=1))***. This measn that state transition P occurs when the allocated period for the traffic lights at ***RED*** is exhausted and there are cars queuing at road ***S1T*** or road ***S3T*** and there are no cars queuing at road ***S2T***, ***S4T***, ***S1***, and ***S3***.
- State transition Q occurs when ***timerRed*** is at logic "1" AND sensor ***M2T***, ***M4T***, ***M1S***, ***M3S***, ***M1T***, and ***M3T*** are at logic "0" AND either sensor ***M2S*** or ***M4S*** is at logic "1" ***(timerRed=1 & M2T=0 & M4T=0 & M1S=0 & M3S=0 & M1T=0 & M3T=0 & (M2S=1 | M4S=1))***. This means that state transition Q occurs when the allocated period for the traffic lights at ***RED*** is exhausted and there are cars queuing at road ***S2*** or road ***S4*** and there are no cars queuing at road ***S2T***, ***S4T***, ***S1***, ***S3***, ***S1T***, or ***S3T***.

12. In ***STATE11***, the traffic lights on road ***S2T*** and road ***S4T*** are ***GREEN***. ***STATE11*** will transition to ***STATE12*** when ***timerGreen*** is at logical "1" (the allocated period for the traffic lights at ***GREEN*** is exhausted).
13. In ***STATE12***, the traffic lights on road ***S2T*** and road ***S4T*** are ***YELLOW***. ***STATE12*** will transition to ***STATE1*** when ***timerYellow*** is at logical "1" (the allocated period for the traffic lights at ***YELLOW*** is exhausted).

Based on the state diagram shown in Figure 4.42 and the interface pins shown in Figure 4.41 and described in Table 4.3, synthesizable Verilog code is written. Example 4.62 shows the Verilog code.

Example 4.62 Synthesizable Verilog Code for Traffic Light State Machine Controller

```
module state_machine (timerGreen, timerYellow,
timerRed, M1S, M2S, M3S, M4S, M1T, M2T, M3T, M4T,
clock, reset, S1_S3, S2_S4, S1T_S3T, S2T_S4T);

input timerGreen, timerYellow, timerRed, M1S, M2S,
M3S, M4S;
input M1T, M2T, M3T, M4T;
input clock, reset;
```

```
parameter [1:0] GREEN = 1,
                YELLOW = 2,
                RED = 3;

output [1:0] S1_S3, S2_S4, S1T_S3T, S2T_S4T;

parameter [3:0] STATE1 = 1,
                STATE2 = 2,
                STATE3 = 3,
                STATE4 = 4,
                STATE5 = 5,
                STATE6 = 6,
                STATE7 = 7,
                STATE8 = 8,
                STATE9 = 9,
                STATE10 = 10,
                STATE11 = 11,
                STATE12 = 12;

reg [3:0] present_state, next_state;

always @ (timerGreen or timerYellow or timerRed or M1S
or M2S or M3S or M4S or M1T or M2T or M3T or M4T)
begin
        case (present_state)
            STATE1:
                begin
                    if (timerRed & (M3S | M1S))
                        next_state = STATE2;
                    else if (timerRed & ~M1S & ~M3S &
                    (M1T | M3T))
                        next_state = STATE5;
                    else if (timerRed & ~M1S & ~M3S &
                    ~M1T & ~M3T & (M2S | M4S))
                        next_state = STATE8;
                    else if (timerRed & ~M1S & ~M3S &
                    ~M1T & ~M3T & ~M2S & ~M4S & (M2T |
                    M4T)) next_state = STATE11;
                    else
                        next_state = STATE1;
                end
            STATE2:
                begin
                    if (timerGreen)
                        next_state = STATE3;
```

```
            else
                next_state = STATE2;
        end
    STATE3:
        begin
            if (timerYellow)
                next_state = STATE4;
            else
                next_state = STATE3;
        end
    STATE4:
        begin
            if (timerRed & ~M1T & ~M3T & ~M2S
            & ~M4S & (M2T | M4T))
                next_state = STATE11;
            else if (timerRed & ~M1T & ~M3T &
            ~M2S & ~M4S & ~M2T & ~M3T & (M1S |
            M3S)) next_state = STATE2;
            else if (timerRed & ~M1T & ~M3T &
            (M2S | M4S))
                next_state = STATE8;
            else if (timerRed & (M1T | M3T))
                next_state = STATE5;
            else
                next_state = STATE4;
        end
    STATE5:
        begin
            if (timerGreen)
                next_state = STATE6;
            else
                next_state = STATE5;
        end
    STATE6:
        begin
            if (timerYellow)
                next_state = STATE7;
            else
                next_state = STATE6;
        end
    STATE7:
        begin
            if (timerRed & ~M2S & ~M4S & ~M2T
            & ~M4T & ~M1S & ~M3S & (M1T |
            M3T)) next_state = STATE5;
```

```
        else if (timerRed & ~M2S & ~M4S &
            ~M2T & ~M4T & (M1S | M3S))
            next_state = STATE2;
        else if (timerRed & (M2S | M4S))
            next_state = STATE8;
        else if (timerRed & ~M2S & ~M4S &
            (M2T | M4T))
            next_state = STATE11;
        else
            next_state = STATE7;
    end
STATE8:
    begin
        if (timerGreen)
            next_state = STATE9;
        else
            next_state = STATE8;
    end
STATE9:
    begin
        if (timerYellow)
            next_state = STATE10;
        else
            next_state = STATE9;
    end
STATE10:
    begin
        if (timerRed & (M2T | M4T))
            next_state = STATE11;
        else if (timerRed & ~M2T & ~M4T &
            (M1S | M3S))
            next_state = STATE2;
        else if (timerRed & ~M2T & ~M4T &
            ~M1S & ~M3S & (M1T | M3T))
            next_state = STATE5;
        else if (timerRed & ~M2T & ~M4T &
        ~M1S & ~M3S & ~M1T & ~M3T &
        (M2S | M4S)) next_state = STATE8;
        else
            next_state = STATE10;
    end
STATE11:
    begin
        if (timerGreen)
            next_state = STATE12;
```

```
                    else
                        next_state = STATE11;
                end
            STATE12:
                begin
                    if (timerYellow)
                        next_state = STATE1;
                    else
                        next_state = STATE12;
                end
            default:
                next_state = STATE1;
        endcase
end

// creation of state flops

always @ (posedge clock or posedge reset)
begin
        if (reset)
                present_state <= STATE1;
        else
                present_state <= next_state;
end

// assignment of output signals

assign S1_S3 = (present_state == STATE2) ? GREEN :
(present_state == STATE3) ? YELLOW : RED;

assign S1T_S3T = (present_state == STATE5) ? GREEN :
(present_state == STATE6) ? YELLOW : RED;

assign S2_S4 = (present_state == STATE8) ? GREEN :
(present_state == STATE9) ? YELLOW : RED;

assign S2T_S4T = (present_state == STATE11) ? GREEN :
(present_state == STATE12) ? YELLOW : RED;

endmodule
```

With the synthesizable Verilog code as shown in Example 4.62, a Verilog test bench is written to simulate the design module ***state_machine***.

Example 4.63 Verilog Test Bench to Simulate Design Module *state_machine*

```
module state_machine_tb ();

parameter [1:0] GREEN = 1,
                YELLOW = 2,
                RED = 3;

reg timerGreen, timerYellow, timerRed;
reg M1S, M2S, M3S, M4S, M1T, M2T, M3T, M4T;
reg clock, reset;
wire [1:0] S1_S3, S2_S4, S1T_S3T, S2T_S4T;

initial
begin
    clock = 0;
    forever #10 clock = ~clock;
end

initial
begin
    timerGreen = 0;
    timerYellow = 0;
    timerRed = 0;
    M1S = 0;
    M2S = 0;
    M3S = 0;
    M4S = 0;
    M1T = 0;
    M2T = 0;
    M3T = 0;
    M4T = 0;
    reset = 0;
    #100;
    reset = 1;
    #100;
    reset = 0;
    #100;
end

initial
begin
    #200; // to wait for reset to be over
    // car going straight from S1 to S3;
```

To initialize all the registers to a zero state.

To reset the design module **state_machine** by forcing **reset** to a logic "1".

```
    M1S = 1;
    #100;
    // car going straight from S4 to S2;
    M1S = 0;
    M4S = 1;
    #200;
    // car turning from S1 to S3 and S2 to S4
    M4S = 0;
    M1T = 1;
    M2T = 1;
    #180;
    // cars from M2T cleared up but M1T still queuing
    M2T = 0;
    #180;
    // car going straight from S2
    M1T = 0;
    M2S = 1;
    #150;
    // no more cars
    M2S = 0;
    #100 $stop;
end

// to force the inputs of timerGreen, timerYellow and
timerRed always @ (reset or S1_S3 or S2_S4 or S1T_S3T
or S2T_S4T)
begin
    if (reset)
        begin
            timerGreen = 0;
            timerYellow = 0;
            timerRed = 0;
        end
    else
        begin
            if ((S1_S3 == GREEN) | (S2_S4 == GREEN) |
                (S1T_S3T == GREEN) | (S2T_S4T == GREEN))
                begin
                    timerRed = 0;
                    #90;
                    timerGreen = 1;
                end
            else if ((S1_S3 == YELLOW) | (S2_S4 ==
                YELLOW) | (S1T_S3T == YELLOW) |
                (S2T_S4T == YELLOW))
```

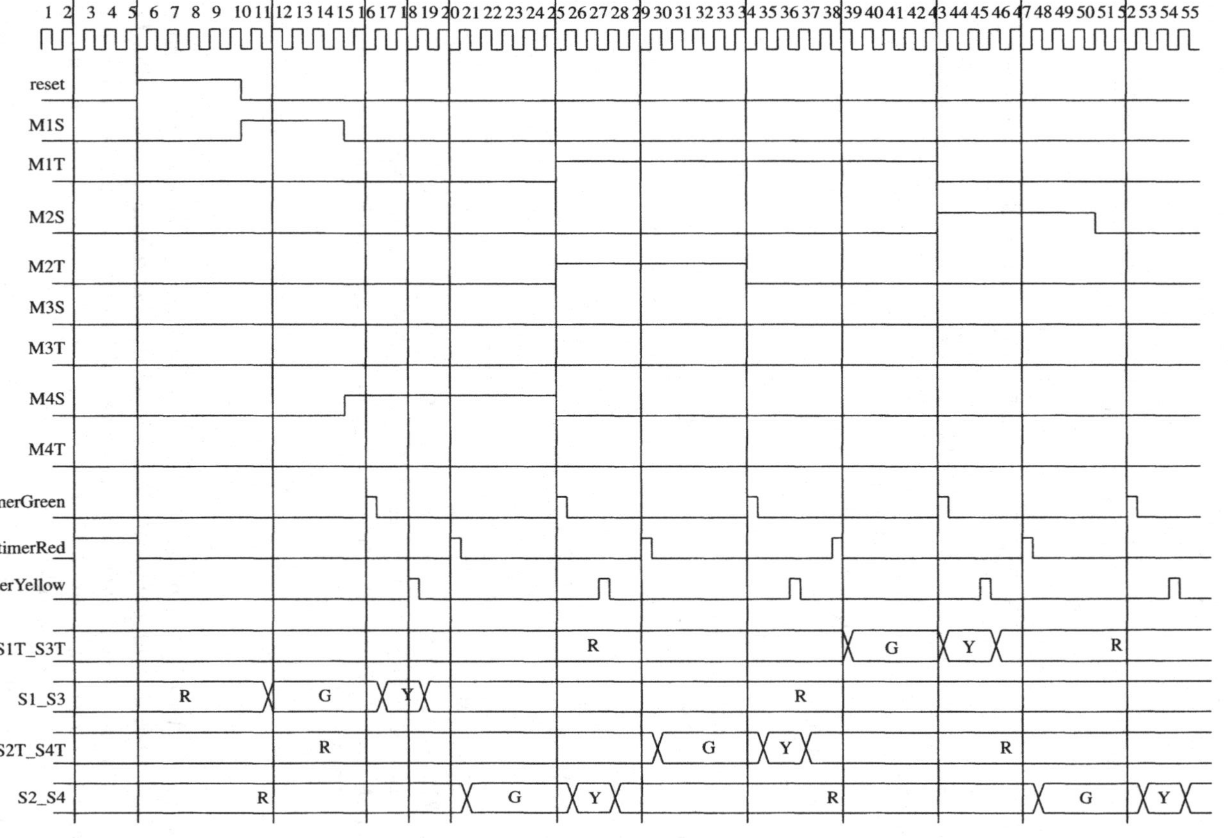

FIGURE 4.43. Diagram showing simulation results of test bench module *state_machine_tb*.

```
                    begin
                        timerGreen = 0;
                        #30;
                        timerYellow = 1;
                    end
                else if ((S1_S3 == RED) | (S2_S4 == RED) |
                    (S1T_S3T == RED) | (S2T_S4T == RED))
                    begin
                        timerYellow = 0;
                        #30;
                        timerRed = 1;
                    end
        end
end

state_machine state_machine_inst (timerGreen,
timerYellow, timerRed, M1S, M2S, M3S, M4S, M1T, M2T,
M3T, M4T, clock, reset, S1_S3, S2_S4, S1T_S3T,
S2T_S4T);

initial
begin
    $monitor ("timerGreen %h timerYellow %h timerRed
%h M1S %b M2S %b M3S %b M4S %b M1T %b M2T %b M3T %b
M4T %b S1_S3 %h S2_S4 %h S1T_S3T %h S2T_S4T
%h",timerGreen, timerYellow, timerRed, M1S, M2S, M3S,
M4S, M1T, M2T, M3T, M4T, S1_S3, S2_S4, S1T_S3T,
S2T_S4T);
end

endmodule
```

Note: The Verilog test bench is written to show the reader how a test bench can be written to simulate the design module ***state_machine****. It can always be written in some other manner that can achieve the same objective. Also take note that the test bench only simulates a small portion of the conditions of the traffic light state machine controller.*

Figure 4.43 shows the simulation waveform result of the test bench ***state_machine_tb***.

CHAPTER FIVE

Design Example of Programmable Timer

Chapter 3 explained the basic concept of Verilog, and Chapter 4 showed some common known coding methods that are used in synthesis. Chapter 5 shows an example of how a real-life practical design can be achieved, beginning from design specification, architectural definition, coding, and verification.

The example discussed in Chapter 5 is a design of a programmable timer. Timers are common design modules in almost all types of system. The design of the programmable timer begins with a design specification for its features and capabilities.

5.1 PROGRAMMABLE TIMER DESIGN SPECIFICATION

The programmable timer is an eight-bit timer that allows three different modes, a one-shot timer, a pulse generator, and a 50% duty cycle waveform generator. For each mode, a certain value can be loaded into the timer before the timer begins clocking. The timer can determine which mode to operate in by using an internal register. This internal register is referred to as "***control word register***." The ***control word register*** is a three-bit register with the MSB bit representing "timer enable" and bit 1 and bit 0 represents the mode of operation.

Referring to Figure 5.1:

Verilog Coding for Logic Synthesis, edited by Weng Fook Lee
ISBN 0-471-429767

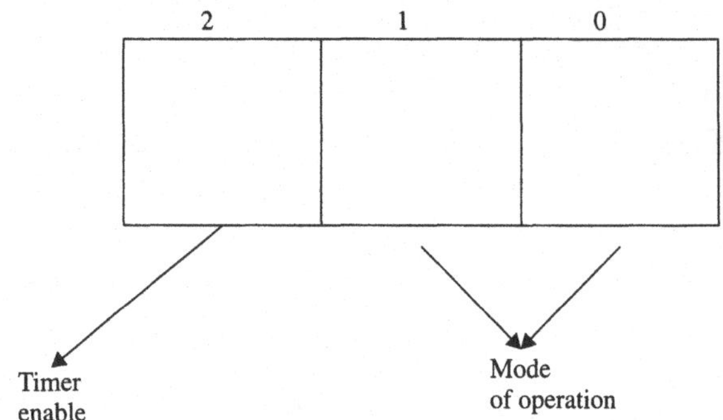

FIGURE 5.1. Diagram showing bits of ***control word register***.

1. Timer enable—this bit if set to a "1" would enable the timer.
2. Mode of operation:
   - Mode "00"—one-shot timer. In this mode, the timer is loaded with an eight-bit binary value. A three-bit value of "100" (timer enable and operate in mode 0) is written into the ***control word register*** of the timer. The timer then begins to count up through each rising edge of clock. When the timer reaches a hex value of "FF", the output of the timer generates a one-clock-width pulse. The timer would then overwrite its MSB of the ***control word register*** to a "0" and the timer would stop. To continue the one-shot timer, the MSB of the ***control word register*** would need to be rewritten to a value of "1".
   - Mode "01"—pulse generator. In this pulse generator mode, the timer is loaded with an eight-bit binary value. A three-bit value of "101" (timer enable and operation in mode 1) is written into the ***control word register*** of the timer. The timer then begins to count up through each rising edge of clock. When the timer reaches a hex value of "FF", the output of the timer would generate a one-clock-width pulse. The timer is then automatically reloaded again with the initial value that was loaded into the timer and the count begins again. Unlike mode 0, the MSB of the ***control word register*** is not overwritten with a "0". The timer in this mode does not stop unless the user writes a "0" to the MSB of the ***control word register*** or if ***ceb*** is pulled to a logic "1".
   - Mode "10"—50% duty cycle waveform generator. In this waveform generator mode, the timer is loaded with an eight-bit binary value. A three-bit value of "110" (timer enable and operation in mode 2) is written into the ***control word register*** of the timer. Unlike mode 0 or 1, the timer would count down and not count up. When the timer reaches a hex value that is half of the loaded value, the output of the timer would generate a logic "1". The timer continues to count down and when it reaches a hexadecimal value of "00", the timer would be

reloaded automatically with the initial loaded value and count down again. Similar to mode 1, the timer in this mode does not stop unless the user writes a "0" to the MSB of the ***control word register*** or if ***ceb*** is pulled to a logic "1".

In mode 2 of operation, the output of the timer would oscillate between 0 and 1 whenever the timer reaches half of the value loaded into the timer. This would create a waveform generator that has 50% duty cycle.

- Mode "11"—not used.

Apart from the ***control word register***, the programmable timer has an eight-bit latch and an eight-bit counter. The eight-bit latch allows a user to latch in a certain binary value. This value is then loaded into the eight-bit counter when the programmable timer is enabled.

With these requirements in mind, an interface specification is defined for the input and output signals.

Table 5.1 shows the signal description for each of the input and output port for the programmable timer design as shown in Figure 5.2.

5.2 MICROARCHITECTURE DEFINITION FOR PROGRAMMABLE TIMER

Based on the design specification from Section 5.1, a microarchitecture can be derived for the design. The requirements of a ***control word register*** would

TABLE 5.1. Signal description for programmable timer design

Signal	Direction	Description
reset	input	***reset*** pin that allows the programmable timer to be ***reset*** to a known state.
ceb	input	chip enable pin. When at logic "0" would enable the programmable timer.
write	input	***write*** pin allows data to be written into the ***control word register*** when the pin is at logic "1". The data at bits 2 to 0 of ***data_in(7:0)*** is written to the three-bit register of ***control word register***.
load	input	***load*** pin allows data at ***data_in(7:0)*** to be loaded into the programmable timer. The data loaded would be the starting point of the count up/down when the programmable timer is enabled.
data_in(7:0)	input	eight-bit bus that is the input for data for the programmable timer.
data_out	output	a one-bit output pin that is the output from the programmable timer.
clk	input	clock input

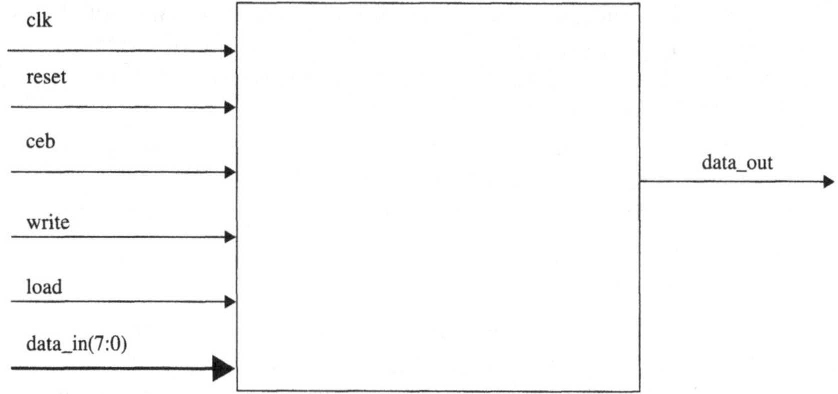

FIGURE 5.2. Diagram showing the interface signals for programmable timer design.

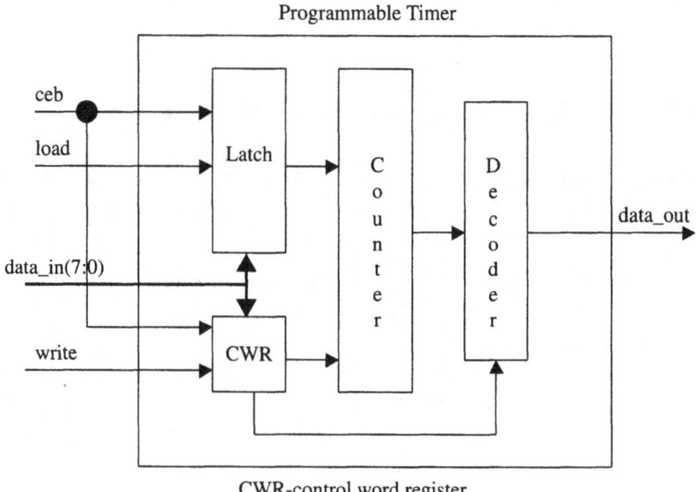

FIGURE 5.3. Microarchitectural block diagram of the programmable timer design.

point to the need of a three-bit register, with each bit being able to control the programmable timer. Because the programmable timer is eight bits wide, there must be an eight-bit counter that is able to count up or count down depending on the mode of execution. A decoder is also needed to decode the output of the eight-bit counter (depending on the mode of execution) to generate an output waveform for the programmable timer. Figure 5.3 shows a microarchitectural block diagram of the programmable timer design.

Referring to Figure 5.3, the programmable timer microarchitecture is partitioned into four major portions:

1. latch. This is an eight-bit latch to latch in the value that is loaded into the counter as the starting point for the counter to count up/down.
2. CWR. This is the ***control word register***, which is a three-bit register that determines if the counter is enabled and what mode to enable the counter in.
3. counter. This is an eight-bit counter that counts up/down depending on which mode the programmable timer is operating in.
4. decoder. This is a decoder that decodes the output of the counter to generate the output for the programmable timer.

5.3 FLOW DIAGRAM DEFINITION FOR PROGRAMMABLE TIMER

Before synthesizable Verilog code is written for the programmable timer, a designer should always spend some time drawing a flow diagram to represent the process flow of the design. Such a diagram can be very useful when the designer starts to write the Verilog code for the design.

The designer does not need to restrict himself/herself to only using flow diagrams. The designer can choose to draw state diagrams, bubble diagrams, or even pseudocode, depending on which is more suitable.

Figure 5.4 shows a flow diagram that covers the conditions that the user can use to load binary data into the programmable timer or writing data into the ***control word register***.

Referring to Figure 5.4:

1. When ***ceb*** and ***load*** are at logic "0" while ***write*** is at logic "1", the data from ***data_in*** bits 2 to 0 are written into the ***control word register***.

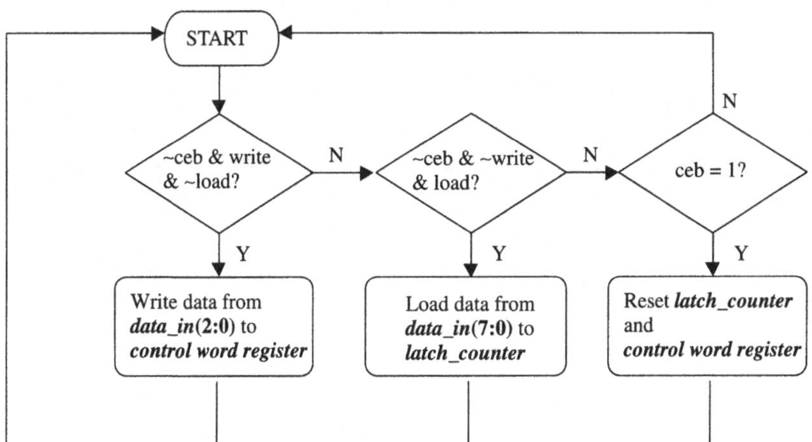

FIGURE 5.4. Flow diagram for loading data into programmable timer and writing data into control word register.

2. When ***ceb*** and ***write*** are at logic "0" while ***load*** is at logic "1", the data from ***data_in*** are written into ***latch_counter***. This is an eight-bit register in the programmable timer that is used to store the value that is loaded into the counter. The counter in the programmable timer will count up or down (depending on mode of operation) using this value in the ***latch_counter*** as the starting count. Furthermore, in mode 1 and mode 2 operation of the programmable timer, automatic reload of the counter is required. Therefore, some form of storage is required for the initial data.
3. When ***ceb*** is at logic "1", ***latch_counter*** and ***control word register*** are reset.

Figure 5.5 shows the flow diagram for resetting the programmable timer when a rising edge is detected at the ***reset*** signal. Notice that during ***reset***, the variables ***flag_counter***, ***counter***, and ***flag_half_counter*** are reset. These variables are flags used in the design of the programmable timer for different modes that it is able to operate in. How these variable are used is shown in the flow diagrams in Figures 5.6, 5.7, and 5.8.

Referring to Figure 5.6:

1. At positive edge of ***clk***, bit 2 of ***control word register*** is checked for a logic value of "1" (logic value "1" on bit 2 indicates the programmable timer is enabled). If it is, then bit 1 and 0 of ***control word register*** is checked for a value of "00" (bits 1 and 0 indicates the mode of opera-

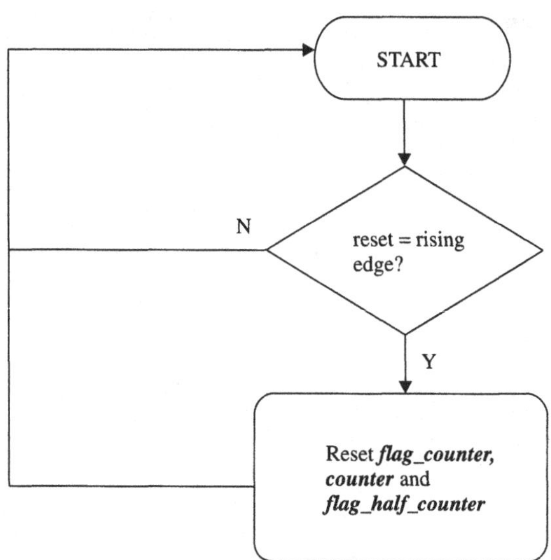

FIGURE 5.5. Flow diagram for reset of programmable timer.

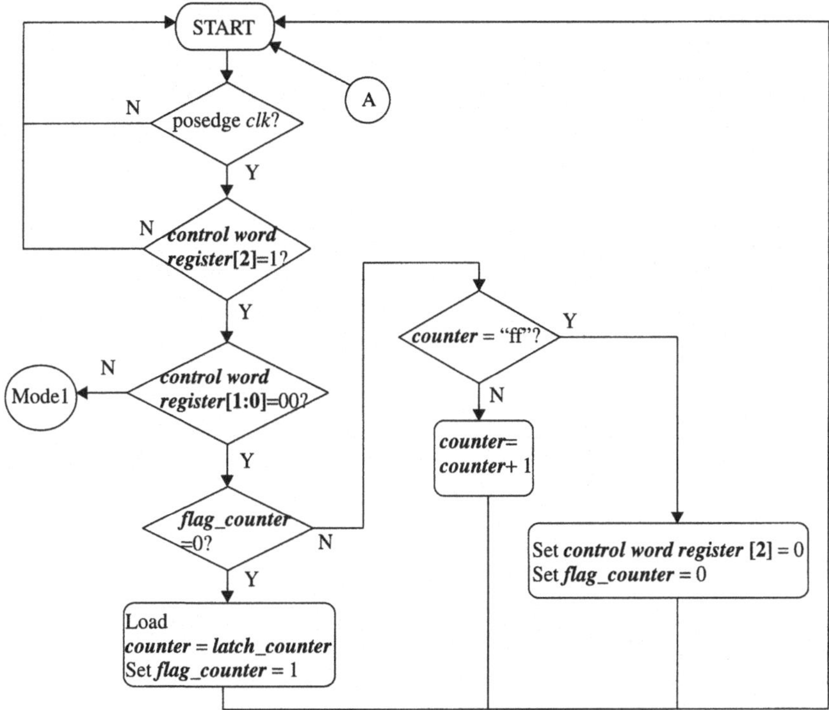

FIGURE 5.6. Flow diagram for mode 0 operation.

tion for the programmable timer). If it is, a flag called ***flag_counter*** is checked for a logic value of "0" (***flag_counter*** is used to indicate that the value in the ***latch_counter*** register is loaded into the counter). If it is, the value in the ***latch_counter*** registers are loaded into the counter and the ***flag_counter*** is set to a logic "1".

2. If ***control word register*** is detected at a value of "100" and ***flag_counter*** is at logic "1", the counter is checked for a value of "ff". If it is, then bit 2 of ***control word register*** is reset and ***flag_counter*** is also reset. If the counter is at a value other than "ff", the counter is incremented by 1.

Figure 5.7 shows a flow diagram (continue from flow diagram of Figure 5.6) for programmable timer at mode 1 execution.

Referring to Figure 5.7:

1. From the connecter for Mode1 from Figure 5.6, bits 1 and 0 of ***control word register*** are checked for logic value of "01" (bits 1 and 0 indicate the mode of operation for the programmable timer). If the logic value is "01", a flag called ***flag_counter*** is checked for a logic value of "0" (***flag_counter*** is used to indicate that the value in the ***latch_counter*** reg-

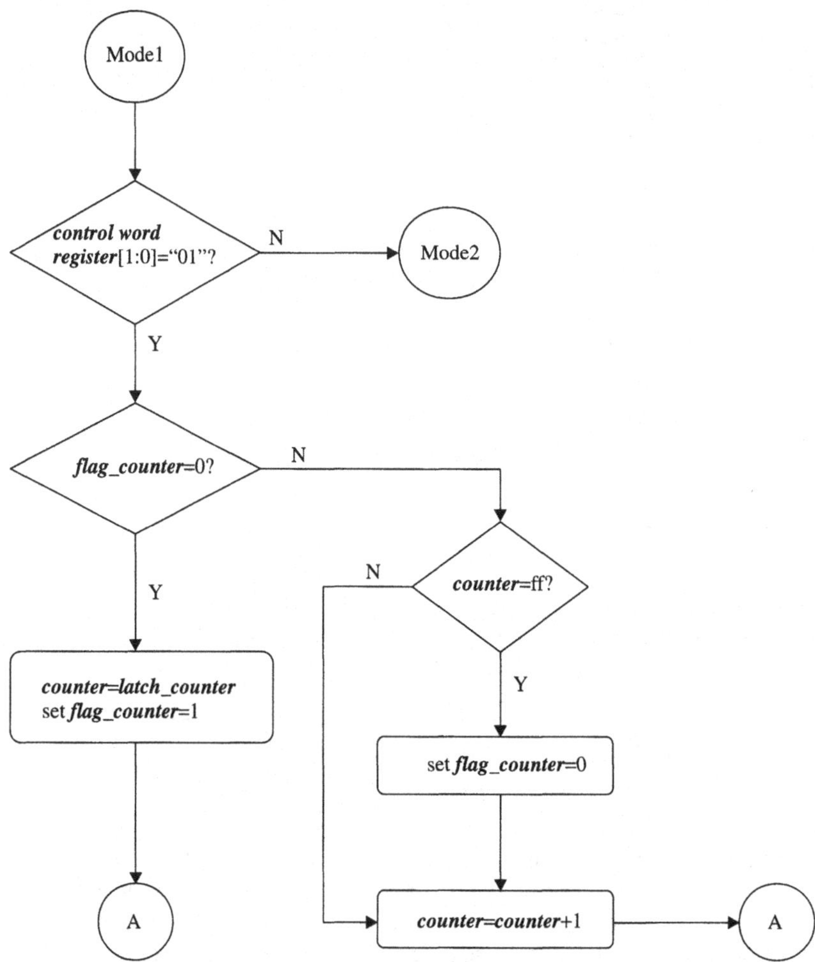

FIGURE 5.7. Flow diagram for mode 1 operation.

ister is loaded into the counter). If it is, the value in the ***latch_counter*** registers are loaded into the counter and ***flag_counter*** is set to a logic "1".

2. If ***control word register*** is detected at a value of "101" and ***flag_counter*** is at logic "1", the counter is checked for a value of "ff". If it is, ***flag_counter*** is reset. The counter would then be incremented by 1. If the counter has a value other than "ff", the counter is incremented by 1.

Figure 5.8 shows a flow diagram (continue from flow diagram of Figure 5.7) for a programmable timer at mode 2 execution.

Referring to Figure 5.8:

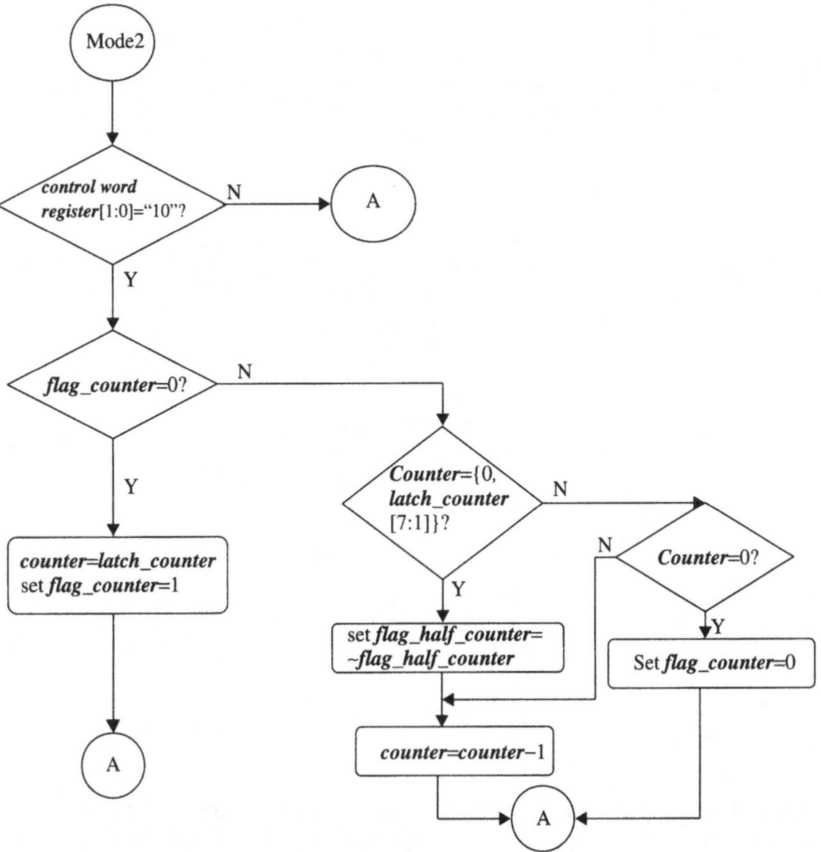

FIGURE 5.8. Flow diagram for mode 2 operation.

1. From the connecter for mode 2 from Figure 5.7, bits 1 and 0 of ***control word register*** are checked for a logic value of "10" (bits 1 and 0 indicate the mode of operation for the programmable timer). If the logic value is "10", a flag called ***flag_counter*** is checked for a logic value of "0" (the ***flag_counter*** is used to indicate that the value in the ***latch_counter*** register has been loaded into the counter). If it is, the value in the ***latch_counter*** registers is loaded into the counter and ***flag_counter*** is set to logic "1".
2. If ***control word register*** is detected at a value of "110" and ***flag_counter*** is at logic "1", the counter is checked for a value that is half of the value stored in ***latch_counter***. If it is, ***flag_half_counter*** is set to the inverse of its previous value. The counter would then be decremented by 1. If the counter has a value other than half of the value stored in ***latch_counter***, the counter is then checked for a value of 0. If it is not 0, then the counter

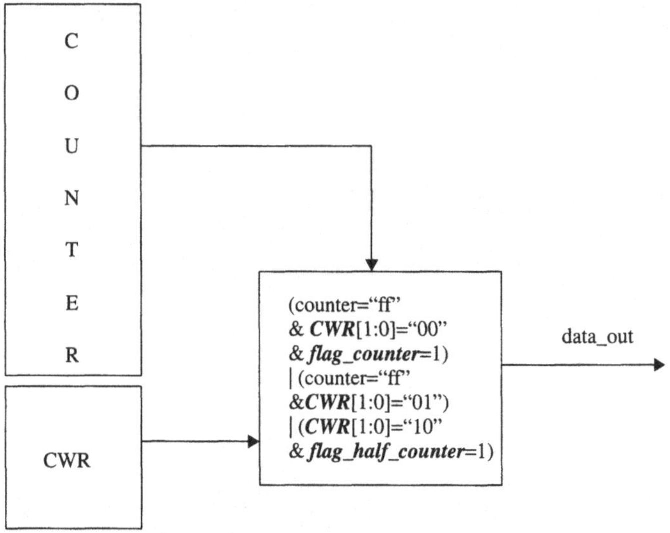

FIGURE 5.9. Diagram showing decoding of control word register and counter for output generation.

would be decremented by 1. However if the counter is at a value of 0, then ***flag_counter*** is reset.

The flow diagrams in Figures 5.6, 5.7, and 5.8 show three different flow diagrams that allow the programmable timer to execute its three different modes. Figure 5.9 shows the conditions of decodings of the counter to create the necessary output for the programmable timer.

Referring to Figure 5.9, decoding to obtain the output ***data_out*** as a logic "1" occurs:

a. when the counter reaches a value of "ff" and bits 1 and 0 of ***control word register*** are decoded as "00" and ***flag_counter*** is "1";
b. when the counter reaches a value of "ff" and bits 1 and 0 of ***control word register*** are decoded as "01";
c. when the ***control word register*** is decoded as "10" and ***flag_half_counter*** is decoded as "1".

Note: Decoding to generate the output **data_out** *uses signal* **flag_counter** *only when the programmable timer is in mode 0 execution. During mode 0 execution (one-shot mode), the programmable timer only creates one pulse and then is disabled. For this reason, the* **data_out** *is decoded with* **flag_counter**, *which is reset to logic "0" when the counter reaches a value of "ff".*

5.4 VERILOG CODE FOR PROGRAMMABLE TIMER

Based on the specifications and flow diagrams shown in Sections 5.1, 5.2 and 5.3, Verilog code is written for the programmable timer (as shown in Example 5.1).

Note: The Verilog code shown in Example 5.1 is not entirely synthesizable. However, the code shown is a simulation that will give the correct results with reference to the requirements of the programmable timer based on the specification in Section 5.1, the microarchitecture definition in Section 5.2, and the flow diagrams in Section 5.3. What this example is trying to point out is that designers need to be careful when writing Verilog code (or VHDL code) for synthesis. Very often, inexperienced designers write code that would simulate to give the right simulation results but would synthesize to some garbage circuits or, even worse, would not synthesize at all. Example 5.1 is one such example. It is a piece of Verilog code that looks correct and simulates correctly but will not synthesize. Why Example 5.1 is not synthesizable is discussed in detail in Section 5.5. Can you detect which part of the code in Example 5.1 is causing the code to be nonsynthesizable? For the time being, let's assume that the code in Example 5.1 is synthesizable.

Example 5.1 Verilog Code for Programmable Timer

```
module prog_counter (
reset, ceb, write, data_in, clk,
load, data_out);

input reset;
input ceb, write, load;
input [7:0] data_in;
input clk;

output data_out;

// to declare the control word
reg [2:0] control_word_register;

// to declare counter with 8 bits
reg [7:0] counter;
reg [7:0] latch_counter;

// flag for first clk pulse after loading in value of
// counter
reg flag_counter;
```

```
// flag for half count cycle
reg flag_half_counter;

// to write control word into counter
// for control_word, bit 2 represent enable, bits 1
// and 0
// represent counter mode
// this also latches in the counter value

always @ (ceb or write or reset or load or data_in)
begin
    if (~ceb & write & ~load & ~reset)
        control_word_register = data_in [2:0];
    else if (~ceb & ~write & load & ~reset)
        latch_counter = data_in;
    else if (ceb & ~reset)
        begin
            // reset the control word counter
            control_word_register = 0;
            // reset the latch_counter
            latch_counter = 0;
        end
end

// to count for counter

always @ (posedge clk or posedge reset)
begin
    if (reset)
    begin
        flag_counter <= 0;
        counter <= 0;
        flag_half_counter <= 0;
    end
    else
    begin
        if (control_word_register[2]) // counter is
        // enabled
        begin
            if (control_word_register[1:0] == 2'b00)
            // this is for one shot mode
            begin
                if (~flag_counter)
                    begin
```

to write to the ***control word register***

to latch data into ***latch_counter***

to reset the ***control word register*** and ***latch_counter***

to reset ***flag_counter***, counter and ***flag_half_counter***

check for ***control word register*** bit 2 to ensure the programmable timer is enabled

check for ***control word register*** for value of "00" for mode 0

Load the value from ***latch_counter*** to counter. Once this is completed, set ***flag_counter*** to a logic "1". ***flag_counter*** is to indicate that the value from ***latch_counter*** has been loaded into the counter.

```
                    counter <= latch_counter;
                    flag_counter <= 1;
                    end
                else
                    begin
                    if (counter == 8'hff)
                        begin
                        // to stop counter for
                        // one shot mode
                        control_word_register[2] <= 0;
                        flag_counter <= 0;
                        end
                    else
                        counter <= counter + 1;
                    end
            end
            else if (control_word_register[1:0] ==
            2'b01) // this is for waveform
            // generator
            begin
                if (~flag_counter)
                    begin
                        counter <= latch_counter;
                        flag_counter <= 1;
                    end
                else
                    begin
                        if (counter == 8'hff)
                            flag_counter <= 0;
                        counter <= counter + 1;
                    end
            end
            else if (control_word_register[1:0] ==
            2'b10) // this is for 50% duty cycle
            // waveform generator
            begin
                if (~flag_counter)
```

When counter reaches value of "ff", reset bit 2 of ***control word register*** to stop the programmable timer and reset the ***flag_counter***.

```
                    begin
                        counter <= latch_counter;
                        flag_counter <= 1;
                    end
                else
                    begin
                        if (counter == {1'b0,
                        latch_counter[7:1]})
                            begin
                                flag_half_counter <=
                                ~flag_half_counter;
                                counter <= counter - 1;
                            end
                        else
                            if (counter == 0)
                                flag_counter <= 0;
                            else
                                counter <= counter - 1;
                    end
            end
        end
    end
end

assign data_out = (
                ((counter == 8'hff) & (control_word_
                register [1:0] == 2'b00) &
                flag_counter) |
                ((counter == 8'hff) & (control_word_
                register [1:0] == 2'b01)) |
                (flag_half_counter & (control_word_
                register [1:0] == 2'b10))
            );

endmodule
```

> to check for value of counter being half of the value stored in ***latch_counter***.

> Decoding to create output ***data_out***.

Example 5.2 shows a test bench that is used to simulate the Verilog code of the programmable timer executing in mode 0.

Example 5.2 Verilog Test Bench to Simulate Programmable Timer Executing in Mode 0

```
module prog_counter_tb ();

reg reset, ceb, write, clk, load;
```

```
reg [7:0] data_in;
wire data_out;

reg [7:0] data;

parameter cycle = 50;

initial
begin
     clk = 0;
     forever #cycle clk = ~clk;
end

initial
begin
     reset = 0;
     ceb = 1;
     write = 0;
     load = 0;
     data_in = 0;
     #cycle
     // setting for reset to 1
     reset = 1;
     repeat (3) #cycle;
     reset = 0;
     #cycle;

     // to set ceb to low to enable chip
     ceb = 0;
     #cycle;

     // load values into the counter

     data = 8'hf0;
     load_counter(data);

     // write values into control word register
     // to enable counter and to put counter in mode 0
     // which is a one shot mode

     data = 8'b00000100;
     write_cwr(data);

     #3000;
     $stop;
```

Declaration of clk

```
end

task write_cwr;
input [7:0] data;
begin
     write = 1;
     data_in = data;
     repeat (2) #cycle;
     write = 0;
     data_in = 0;
     #cycle;
end
endtask

task load_counter;
input [7:0] data;
begin
     load = 1;
     data_in = data;
     repeat (2) #cycle;
     load = 0;
     data_in = 0;
     #cycle;
end
endtask

prog_counter prog_counter_inst (reset, ceb, write,
data_in, clk, load, data_out);

endmodule
```

Figure 5.10 shows the simulation waveform for the Verilog test bench used to simulate the programmable timer at mode 0 execution.

Referring to Figure 5.10:

1. In clock 1 and clock 2, ***reset*** signal is at logic "1". This resets the counter, ***flag_counter***, and ***flag_half_counter***.
2. In clock 3, ***ceb*** goes to logic "0" to enable the programmable timer.
3. In clock 4, ***load*** signal is at logic "1". This would load the value "f0" at the ***data_in*** bus to the ***latch_counter***.
4. In clock 5, ***write*** signal is at logic "1". This will write the value "4" at the ***data_in*** bus to the ***control word register***.
5. In clock 6, ***write*** and ***load*** is at logic "0", while ***control word register*** is at a value of "100". Bits 1 and 0 decode the operation of programmable timer in mode 0. Bit 2 decodes the enabling of the programmable timer.

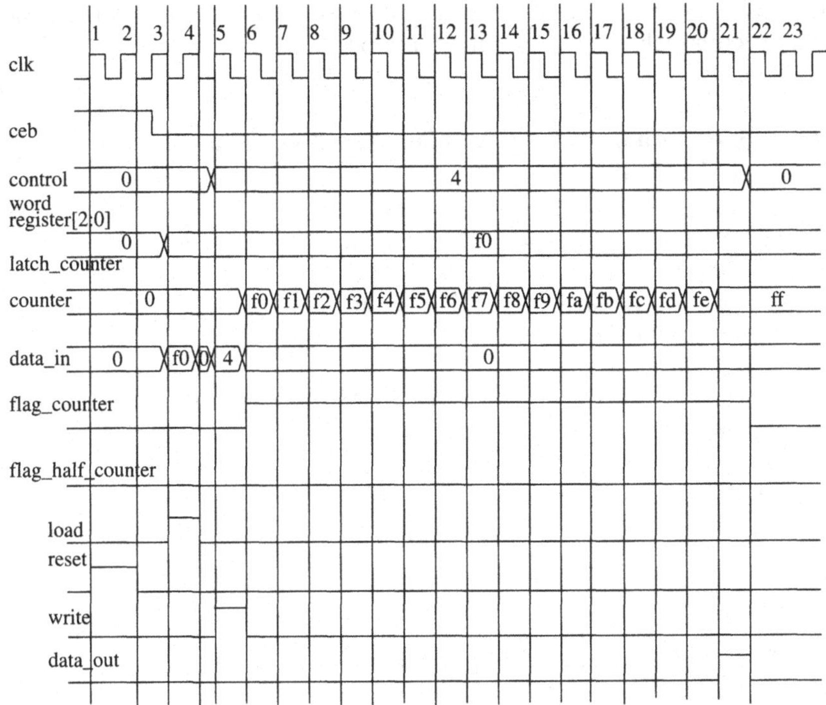

FIGURE 5.10. Diagram showing simulation results of Verilog test bench for mode 0.

Therefore, at clock 6, the value of ***latch_counter*** is loaded into the counter, thereby having the counter at a value of "f0". During clock 6, the signal ***flag_counter*** is driven to logic "1".

6. From clock 7 to clock 20, the counter would increment by 1 during every rising edge of clock.
7. In clock 21, the counter reaches a value of "ff". This is decoded to create an output on ***data_out*** with a logic "1".
8. In clock 22, bit 2 of ***control word register*** is reset to a logic "0". This would disable the programmable timer. At the same time ***flag_counter*** is also reset.

*Note: The signal **flag_counter** is created as a flag to signify that the counter has already been loaded with the value from the **latch_counter**. This flag is then used together with the counter value and **control word register** bits 1 and 0 to decode for the output **data_out**.*

*Also, note that in mode 0, the programmable timer functions as a one-shot device. Once the output **data_out** drives a pulse, it would then stop. The programmable timer will need to be enabled again by writing a command to the **control word register** to enable the programmable timer.*

Example 5.3 shows a test bench that is used to simulate the Verilog code of the programmable timer executing in mode 1.

Example 5.3 Verilog Test Bench to Simulate Programmable Timer Executing in Mode 1

```
module prog_counter_tb ();

reg reset, ceb, write, clk, load;
reg [7:0] data_in;
wire data_out;

reg [7:0] data;

parameter cycle = 50;

initial
begin
     clk = 0;
     forever #cycle clk = ~clk;
end

initial
begin
     reset = 0;
     ceb = 1;
     write = 0;
     load = 0;
     data_in = 0;
     #cycle
     // setting for reset to 1
     reset = 1;
     repeat (3) #cycle;
     reset = 0;
     #cycle;

     // to set ceb to low to enable chip
     ceb = 0;
     #cycle;

     // load values into the counter

     data = 8'hfa;
     load_counter(data);
```

```
    // write values into control word register
    // to enable counter and to put counter in mode 1
    // which is a pulse waveform generator

    data = 8'b00000101;
    write_cwr(data);

    #3000;
    $stop;
end

task write_cwr;
input [7:0] data;
begin
    write = 1;
    data_in = data;
    repeat (2) #cycle;
    write = 0;
    data_in = 0;
    #cycle;
end
endtask

task load_counter;
input [7:0] data;
begin
    load = 1;
    data_in = data;
    repeat (2) #cycle;
    load = 0;
    data_in = 0;
    #cycle;
end
endtask

prog_counter prog_counter_inst (reset, ceb, write,
data_in, clk, load, data_out);

endmodule
```

Figure 5.11 shows the simulation waveform for the Verilog test bench used to simulate the programmable timer at mode 1 execution.

Referring to Figure 5.11:

1. In clock 1 and clock 2, ***reset*** signal is at logic "1". This resets the counter, ***flag_counter***, and ***flag_half_counter***.

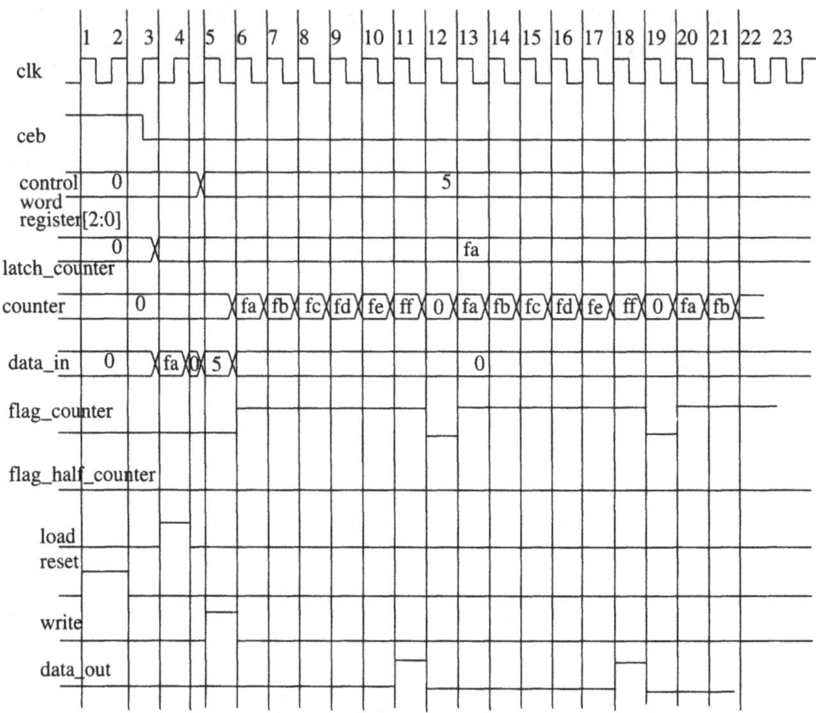

FIGURE 5.11. Diagram showing simulation results of Verilog test bench for mode 1.

2. In clock 3, ***ceb*** goes to logic "0" to enable the programmable timer.
3. In clock 4, ***load*** signal is at logic "1". This would load the value "fa" at the ***data_in*** bus to the ***latch_counter.***
4. In clock 5, ***write*** signal is at logic "1". This will write the value "5" at the ***data_in*** bus to the ***control word register.***
5. In clock 6, ***write*** and ***load*** is at logic "0", while ***control word register*** is at a value of "101". Bits 1 and 0 decode the operation of the programmable timer in mode 1. Bit 2 decodes to enabling of programmable timer. At clock 6, the value of ***latch_counter*** is loaded into the counter, thereby having the counter at a value of "fa". During clock 6, the signal ***flag_counter*** is driven to logic "1".
6. From clock 7 to clock 11, the counter would increment by 1 during every rising edge of clock.
7. At clock 11, the counter reaches the value of "ff". This counter value is decoded to create the output ***data_out*** to be at logic "1". During this clock, the signal ***flag_counter*** is reset to 0. However, the ***flag_counter*** does not go to a logic "0" immediately in clock 11, but instead would go to a logic "0" in clock 12. This occurs because in clock 11, when the

counter is at the value of "ff", the ***flag_counter*** is assigned the value of logic "0". And this assignment will take place on the next clock (the signal is flopped).

8. Clock 12 have the ***flag_counter*** at a logic value of "0".
9. In clock 13, the counter is automatically reloaded with the value "fa" in the ***latch_counter***. Signal ***flag_counter*** is at logic value of "1" and the steps from clock 7 is repeated. This will continue until:
   a. a reset happens
   b. ceb goes back to logic "1"
   c. a new value is written into the ***control word register***

Note: For mode 1 of operation, the timer automatically reloads the counter everytime ***flag_counter*** *switches from logic "0" to logic "1". Therefore, the output waveform at* ***data_out*** *is a pulse generator.*

Example 5.4 shows a test bench that is used to simulate the Verilog code of the programmable timer executing in mode 2.

Example 5.4 Verilog Test Bench to Simulate Programmable Timer Executing in Mode 2

```
module prog_counter_tb ();

reg reset, ceb, write, clk, load;
reg [7:0] data_in;
wire data_out;

reg [7:0] data;

parameter cycle = 50;

initial
begin
     clk = 0;
     forever #cycle clk = ~clk;
end

initial
begin
     reset = 0;
     ceb = 1;
     write = 0;
     load = 0;
     data_in = 0;
     #cycle
```

```
    // setting for reset to 1
    reset = 1;
    repeat (3) #cycle;
    reset = 0;
    #cycle;

    // to set ceb to low to enable chip
    ceb = 0;
    #cycle;

    // load values into the counter

    data = 8'h05;
    load_counter(data);

    // write values into control word register
    // to enable counter and to put counter in mode 2
    // which is a 50% duty cycle waveform generator

    data = 8'b00000110;
    write_cwr(data);

    #3000;
    $stop;
end

task write_cwr;
input [7:0] data;
begin
    write = 1;
    data_in = data;
    repeat (2) #cycle;
    write = 0;
    data_in = 0;
    #cycle;
end
endtask

task load_counter;
input [7:0] data;
begin
    load = 1;
    data_in = data;
    repeat (2) #cycle;
    load = 0;
```

```
        data_in = 0;
        #cycle;
end
endtask

prog_counter prog_counter_inst (reset, ceb, write,
data_in, clk, load, data_out);

endmodule
```

Figure 5.12 shows the simulation waveform for the Verilog test bench used to simulate the programmable timer at mode 2 execution.

Referring to Figure 5.12:

1. In clock 1 and clock 2, ***reset*** signal is at logic "1". This resets the counter, ***flag_counter***, and ***flag_half_counter***.
2. In clock 3, ***ceb*** goes to logic "0" to enable the programmable timer.
3. In clock 4, ***load*** signal is at logic "1". This would load the value "5" at the ***data_in*** bus to the ***latch_counter***.
4. In clock 5, ***write*** signal is at logic "1". This will write the value "6" at the ***data_in*** bus to the ***control word register***.
5. In clock 6, ***write*** and ***load*** is at logic "0", while ***control word register*** is at a value of "110". Bits 1 and 0 decode the operation of program-

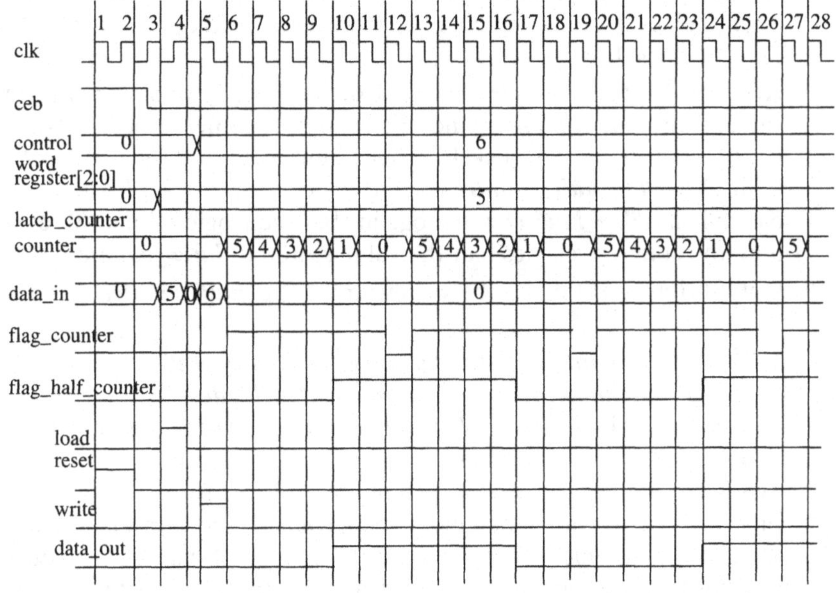

FIGURE 5.12. Diagram showing simulation results of Verilog test bench for mode 2.

mable timer in mode 2. Bit 2 decodes the enabling of the programmable timer. At clock 6, the value of ***latch_counter*** is loaded into the counter, thereby having the counter at a value of "5". During clock 6, the signal ***flag_counter*** is driven to logic "1" to indicate that the values in the ***latch_counter*** have been loaded into the counter.

6. From clock 7 to clock 11, the counter would decrement by 1 during every rising edge of clock.
7. In clock 9, the counter reaches a value of "2". Because the counter was initially loaded with the value 5, half of 5 is 2 (rounded from 2.5). Therefore in clock 9, the ***flag_half_counter*** is assigned to a logic value of "1". This assignment value on ***flag_half_counter*** is not seen at clock 9 but only seen at clock 10. The reason for this is again because in clock 9, when the assignment of logic "1" to ***flag_half_counter*** occurs, the rising edge of clock 9 has already occurred. On the rising edge of clock 9, the counter is decremented by one to a value of "2". The decoding of this value "2" for the counter causes the assignment of logic "1" to ***flag_half_counter***, but the rising edge of clock 9 has already occurred. Therefore, ***flag_half_counter*** would only have the value of logic "1" on the rising edge of clock 10.
8. At clock 11, the counter reaches the value of "00". When this occurs, ***flag_counter*** is reset. Again, similarly as in clock 9, assignment of logic "0" to ***flag_counter*** is not seen until clock 12. This is due to the fact that when the counter is decremented to a value of "00" (during rising edge of clock 11), the decoding of this counter value causes the assignment of logic "0" to ***flag_counter***. However, because it is no longer the rising edge of clock 11, the assignment of the value would occur on the rising edge of clock 12.
9. At clock 13, ***flag_counter*** reset and the counter is reloaded with the value from ***latch_counter*** and the whole counting process repeats. However, note that in clock 16, when counter reaches the value of "2" (half of 5), ***flag_half_counter*** goes to logic "0" in clock 17. The design basically has ***flag_half_counter*** toggling between logic "1" and "0" every time the counter reaches half the value stored in ***latch_counter***.
10. Output ***data_out*** follows the waveform of ***flag_half_counter***.

Note: Mode 2 of operation is similar to mode 1 of operation whereby the counter in the programmable timer is automatically reloaded when the counter reaches a value of "00". This would create an output waveform ***data_out*** *that is a 50% duty cycle waveform generator.*

The synthesizable Verilog code shown in Example 5.1 for the programmable timer has the assignment statements ***flag_counter*** and ***flag_half_counter*** within the ***always*** statement that detects a rising edge of clock. This method of coding flops both the signals. As a result, assignment of values to the signal ***flag_counter*** and ***flag_half_counter*** always happens one clock later (because it is flopped). A designer should always note that when a signal is assigned a value within an ***always*** statement that detects for a rising edge or falling edge, that signal will be flopped.

5.5 SYNTHESIZABLE VERILOG CODE FOR PROGRAMMABLE TIMER

Based on the simulation waveform results shown in Figures 5.10, 5.11 and 5.12, the Verilog code of Example 5.1 gives the correct and expected simulation results. However, when the Verilog code is synthesized, an error will occur. Do you know where the error is?

Referring to the Verilog code in Example 5.1, there are two ***always*** statements. The first ***always*** statement is

```
always @ (ceb or write or reset or load)
begin
     if (~ceb & write & ~load & ~reset)
          control_word_register = data_in [2:0];
     else if (~ceb & ~write & load & ~reset)
          latch_counter = data_in;
     else if (ceb & ~reset)
          begin
               // reset the control word counter
               control_word_register = 0;
               // reset the latch_counter
               latch_counter = 0;
          end
end
```

which creates a set of latches for ***latch_counter*** and ***control_word register***. This is the statement that sets the assignments for the signal ***control_word_register*** as well as ***latch_counter***.

The second ***always*** statement is

```
always @ (posedge clk or posedge reset)
begin
     if (reset)
     begin
          flag_counter <= 0;
          counter <= 0;
          flag_half_counter <= 0;
     end
     else
     begin
```

```
            if (control_word_register[2]) // counter is enabled
            begin
                 if (control_word_register[1:0] == 2'b00)
                 // this is
                 // for one shot mode
                 begin
                      if (~flag_counter)
                           begin
                           counter <= latch_counter;
                           flag_counter <= 1;
                           end
                      else
                           if (counter == 8'hff)
                              begin
                              // to stop counter for one shot
                              // mode
                              control_word_register[2] <= 0;
                              flag_counter <= 0;
                              end
                           else
. . .
. . .
. . .
            end
       end
end
```

which creates a set of flip-flops triggered by rising edge of ***clk*** for signals ***flag_counter***, ***flag_half_counter***, and ***counter***. However, in this second ***always*** statement, during the execution of the programmable timer at mode 0, assignment of value to the ***control word register*** occurs to disable the timer when the one shot pulse has been generated by the timer:

```
control_word_register[2] <= 0;
```

This causes a conflicting problem in the sense that the first ***always*** statement drives a value on the ***control word register*** while the second ***always*** statement also drives a value on the ***control word register***. During synthesis, the synthesis tool is confused and would not know which circuit to synthesize. There are multiple drivers on the node and the synthesis tool would not know which driver is allowed to drive the node (as contention occurs). If the design is coded in such a way as to create the ***control word register*** as a tri-state register, it is possible to have different circuits to drive a node. However, in this case, the ***control word register*** is a logic register that cannot have two separate circuits driving a node.

In synthesis, a synthesis tool would need to know the logic that is required to drive a node before being able to synthesize the necessary circuit to drive a node.

There are many ways to fix this problem. The Verilog code in Example 5.5 shows one method of coding to resolve this problem.

Example 5.5 Synthesizable Verilog Code for Programmable Timer

```
module prog_counter1 (
reset, ceb, write, data_in, clk,
load, data_out);

input reset;
input ceb, write, load;
input [7:0] data_in;
input clk;

output data_out;

// to declare the control word
reg [2:0] control_word_register;
reg disable_CWR;

// to declare counter with 8 bits
reg [7:0] counter;
reg [7:0] latch_counter;

// flag for first clk pulse after loading in value of
// counter
reg flag_counter;
// flag for half count cycle
reg flag_half_counter;

// to write control word into counter
// for control_word, bit 2 represents enable, bit 1 and 0
// represent counter mode
// this also latches in the counter value

always @ (ceb or write or reset or load or disable_CWR
or data_in)
begin
     if (~ceb & write & ~load & ~reset)
          control_word_register = data_in [2:0];
     else if (~ceb & ~write & load & ~reset)
          latch_counter = data_in;
```

Declaration of signal ***disable_CWR***

```
    else if (ceb & ~reset)
        begin
            // reset the control word counter
            control_word_register = 0;
            // reset the latch_counter
            latch_counter = 0;
        end
    else if (disable_CWR)
        control_word_register[2] = 0;
end

// to count for counter

always @ (posedge clk or posedge reset)
begin
    if (reset)
    begin
        disable_CWR <= 0;
        flag_counter <= 0;
        counter <= 0;
        flag_half_counter <= 0;
    end
    else
    begin
        if (control_word_register[2]) // counter is
        // enabled
        begin
            if (control_word_register[1:0] == 2'b00)
            // this is
            // for one shot mode
            begin
                if (~flag_counter)
                    begin
                    counter <= latch_counter;
                    flag_counter <= 1;
                    end
                else
                    if (counter == 8'hff)
                      begin
                      // to stop counter for one shot
                      // mode
                      disable_CWR <= 1;
                      flag_counter <= 0;
                      end
                    else
```

Additional branch for ***if*** statement for condition whereby ***disable_CWR*** is at logic "1" will cause assignment of bit 2 of ***control word register*** to a logic "0".

To reset ***disable_CWR*** during reset.

Setting of ***disable_CWR*** to logic "1" in order to stop the programmable timer.

```
                    counter <= counter + 1;
    end
    else if (control_word_register[1:0] ==
    2'b01) // this
    // is for waveform generator
    begin
         if (~flag_counter)
              begin
                   counter <= latch_counter;
                   flag_counter <= 1;
              end
         else
              begin
                   if (counter == 8'hff)
                   flag_counter <= 0;
                   counter <= counter + 1;
              end
    end
    else if (control_word_register[1:0] ==
    2'b10) // this
    // is for 50% duty cycle waveform
    // generator
    begin
         if (~flag_counter)
              begin
                   counter <= latch_counter;
                   flag_counter <= 1;
              end
         else
              begin
                   if (counter == {1'b0,
                   latch_counter[7:1]})
                        begin
                           flag_half_counter <=
                           ~flag_half_counter;
                           counter <= counter
                              - 1;
                        end
                   else
                        if (counter == 0)
                        flag_counter <= 0;
                        else
                           counter <= counter
                           - 1;
              end
```

```
            end
        end
        else
            begin
                disable_CWR <= 0;
                flag_counter <= 0;
                flag_half_counter
                   <= 0;
            end
    end
end
```

To clear ***disable_CWR*** after programmable timer disabled. This is needed so that a new command can be issued to the ***control word register*** to re-enable the programmable timer.

```
assign data_out = (
                  ((counter == 8'hff) & (control_word_
                      register [1:0] ==
                      2'b00) & flag_counter) |
                  ((counter == 8'hff) & (control_word_
                      register [1:0] == 2'b01)) |
                  (flag_half_counter & (control_word_
                      register [1:0] == 2'b10)) );

endmodule
```

The Verilog code in Example 5.5 is different from Example 5.1 in that a new signal called ***disable_CWR*** is introduced. In Example 5.5, this signal is used to assign a logic "0" to the most significant bit of ***control word register***. In other words, the signal ***disable_CWR*** acts as a qualifier signal for enabling or disabling the MSB of ***control word register***. By so doing, the driver for ***control word register*** is limited to only the first ***always*** statement. And, therefore, there is no longer a situation in which contention may happen.

The Verilog code in Example 5.5 can be simulated using the same test bench in Example 5.2, 5.3, and 5.4. The simulation results are the same as that for the simulation of Verilog code for Example 5.1.

This comparison between Verilog code of Example 5.1 and 5.5 shows a very important difference between writing Verilog code for simulation and for synthesis. A piece of code that simulates correctly does not necessarily mean that it would synthesize. Therefore, it is important for a designer to understand the limitations of writing code for synthesizability.

CHAPTER SIX

Design Example of Programmable Logic Block for Peripheral Interface

Chapter 6 shows an example of designing a programmable logic block for peripheral interface (similar to the industry standard of 8255 PPI), beginning from design specification, architectural definition, coding, and verification. However, please note that the example presented here is very similar to the industry's widely used 8255 PPI, but not fully compatible. The objective of this chapter is to show the reader how a logic module can be designed using synthesizable Verilog. It is not the objective of this example to replace the current industry's 8255 PPI. This example is referred to as programmable logic block for peripheral interface or, in short, "PLB".

A peripheral interface is used widely in electronic systems that require a communication connection between several devices. For example, 8255 PPI (a peripheral interface) is used in a computer to allow for a communication connection between the microprocessor (central processing unit, or CPU) and the peripheral components of the computer.

The design of the PLB begins with a design specification on its features and capabilities.

Note: There are many ways and different styles to code synthesizable Verilog. The example of Verilog code shown in Chapter 6 is written specifically to show different ways that can be used to code certain implementation. Although some

Verilog Coding for Logic Synthesis, edited by Weng Fook Lee
ISBN 0-471-429767

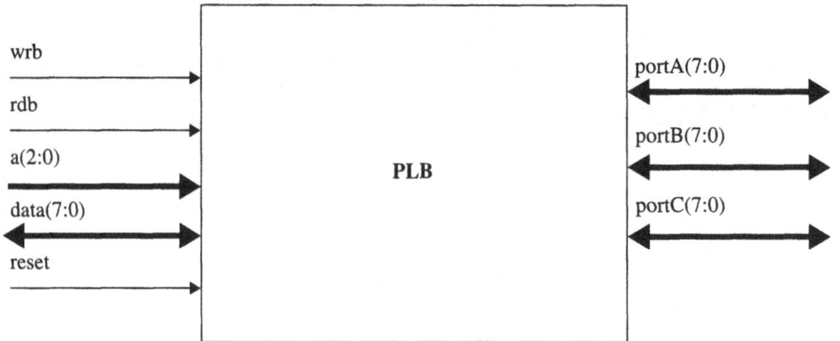

FIGURE 6.1. Diagram showing interface signals of PLB.

logic in this example may be similar, the style of coding written to generate this logic may be different.

6.1 PROGRAMMABLE LOGIC BLOCK FOR PERIPHERAL INTERFACE DESIGN SPECIFICATION

The PLB is an interconnect device that has 24 I/O pins, separated into three groups: ***portA***, ***portB***, and ***portC***. Each of these I/O ports are eight bits wide. ***PortC*** can be further separated into two subgroups, portCupper and portClower, each four bits wide. The PLB can also operate in three different modes: mode 0, mode 1, and mode 2. Each mode of operation allows for different configurations on the I/O ports. Figure 6.1 shows an interface diagram of the PLB and Table 6.1 shows a description of the interface signals.

Internal to the PLB, there are two registers (***CWR*** and ***STATUS***), which are both eight bits wide. ***CWR*** register is used as a control register to determine the mode of functionality and direction of each of ***portA***, ***portB***, and ***portC***. Figure 6.2 shows the function of each bit in the ***CWR*** register.

Referring to Figure 6.2, each bit in the CWR register has its functionality.

1. Bit 7—active bit. This bit must be a logic "1" in order for the PLB to function. A logic "0" on this bit is similar to power down of the chip.
2. Bits 6 and 5—mode of operation.
   a. "00"—Mode 0 operation. In this mode, ***portA***, ***portB***, and ***portC*** can either be an input bus or output bus, depending on bit 4 to bit 1 of ***CWR*** register. Table 6.2 shows a description on the functionality of ***portA***, ***portB*** and ***portC*** based on the logic values of bits 4 to bits 1 of ***CWR*** register.
   b. "01"—Mode 1 operation. In this mode, ***portA*** and ***portB*** can either be a strobed input or strobed output depending on bit 2 and bit 1.

TABLE 6.1. Table showing a description of PLB's interface signals

Signal	Direction	Description
rdb	input	Active low to indicate a read from either ***portA***, ***portB***, or ***portC*** to ***data*** bus.
wrb	input	Active low to indicate a write to either ***portA***, ***portB***, or ***portC*** from ***data*** bus.
reset	input	Active high to reset the PLB.
data(7:0)	I/O	Bidirectional bus for reading of data from ***portA***, ***portB***, or ***portC*** and for writing of data to ***portA***, ***portB***, or ***portC***.
a(2:0)	input	Address signals to identify which port or register the data is meant for: "000"—***portA*** "001"—***portB*** "010"—***portC*** "011"—***CWR*** register "111"—***STATUS*** register
portA(7:0)	I/O	Bidirectional bus of ***portA***
portB(7:0)	I/O	Bidirectional bus of ***portB***
portC(7:0)	I/O	Bidirectional bus of ***portC***

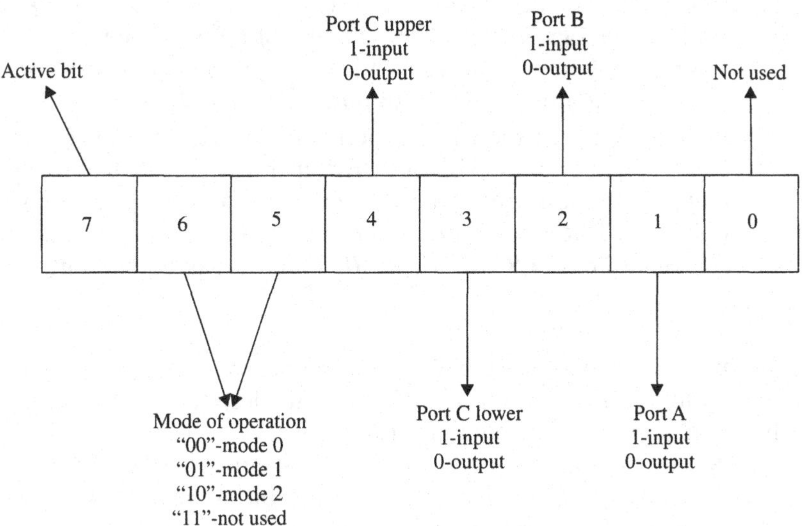

FIGURE 6.2. Diagram showing functionality of each bit in CWR register.

TABLE 6.2. Table showing a description of bits 4 to 1 of *CWR* register

Bit	Logic Value	Functionality of Ports in Mode 0 Operation
4	0	***PortC[7:4]*** operates as output bus
	1	***PortC[7:4]*** operates as input bus
3	0	***PortC[3:0]*** operates as output bus
	1	***PortC[3:0]*** operates as input bus
2	0	***PortB*** operates as output bus
	1	***PortB*** operates as input bus
1	0	***PortA*** operates as output bus
	1	***PortA*** operates as input bus

TABLE 6.3. Table showing a description of bits 2 and 1 of *CWR* register

Bit	Logic Value	Functionality of Ports in Mode 1 Operation
2	0	***PortB*** operates as strobed output bus
	1	***PortB*** operates as strobed input bus
1	0	***PortA*** operates as strobed output bus
	1	***PortA*** operates as strobed input bus

Table 6.3 shows a description on the functionality of ***portA*** and ***portB*** based on the logic values of bits 2 and 1 of ***CWR*** register. Note that in mode 1 operation, ***portC*** is used as the control signals for the strobed input/output function of ***portA*** and ***portB***.

c. "10"—Mode 2 operation. In this mode, ***portA*** is a strobed I/O port, with ***portC*** being the control signals. ***PortB*** operates in mode 0 as either an input port or output port depending on bit 2 of ***CWR***. If the logic value of bit 2 of ***CWR*** is a "1", ***portB*** operates as an input port. If the logic value of bit 2 of ***CWR*** is a "0", ***portB*** operates as an output port.

3. Bit 0—Not used

Data can be written to the ***CWR*** register by pulsing ***wrb*** signal low while having the address (***a2***, ***a1***, ***a0***) at "011", and the desired data to be written into the ***CWR*** register at the data pins (***data[7:0]***).

STATUS register is an eight-bit register that controls output ***portC*** when the PLB is operating in mode 1 or mode 2. Each bit in the ***STATUS*** register controls a separate output bit of ***portC*** in mode 1 and mode 2 of operation. Figure 6.3 shows the function of each bit in the ***STATUS*** register.

Data can be written to the ***STATUS*** register by pulsing ***wrb*** signal low while having the address (***a2***, ***a1***, ***a0***) at "111", and the desired data to be written into the ***STATUS*** register at the data pins (***data[7:0]***).

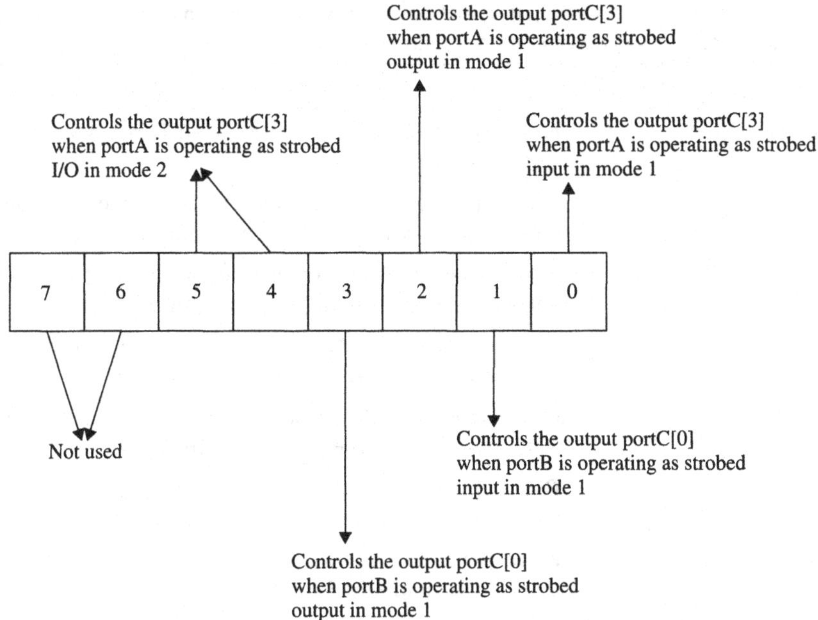

FIGURE 6.3. Diagram showing functionality of each bit in ***STATUS*** register.

Section 6.2 details the operation of the PLB based on the logic values of ***CWR*** register and ***STATUS*** register.

6.2 MODE OF OPERATION FOR PROGRAMMABLE LOGIC BLOCK FOR PERIPHERAL INTERFACE

The PLB can operate in three different modes of operation. The logic values in the ***CWR*** register determines which mode of operation the PLB is to operate in while the ***STATUS*** register determines several output values of ***portC*** during mode 1 and mode 2.

Section 6.2.1 details the operation of the PLB in mode 0, Section 6.2.2 details the operation of the PLB in mode 1, and Section 6.2.3 details the operation of the PLB in mode 2.

6.2.1 Mode 0 Operation

In mode 0 operation, the PLB can be configured using the ***CWR*** register to have ***portA*** as either an input or output port, ***portB*** as either an input or output port, ***portC*** upper (bits 7 to bits 4) as either an input or output port and, ***portC*** lower (bits 3 to bits 0) as either an input or output port. Table 6.4 shows the

TABLE 6.4. Table showing the combinations of functionality of *portA, portB, portC* upper, and *portC* lower in mode 0 operation

CWR Register Values	Description
1000000X	***PortC*** upper, ***PortC*** lower, ***PortB***, and ***PortA*** operate as output ports.
1000001X	***PortC*** upper, ***PortC*** lower, and ***PortB*** operate as output ports while ***portA*** operates as an input port.
1000010X	***PortC*** upper, ***portC*** lower, and ***portA*** operate as output ports while ***portB*** operates as an input port.
1000011X	***PortC*** upper and ***portC*** lower operate as output ports while ***portB*** and ***portA*** operate as input ports.
1000100X	***PortC*** upper, ***portB***, and ***portA*** operate as output ports while ***portC*** lower operates as an input port.
1000101X	***PortC*** upper and ***portB*** operate as output ports while ***portC*** lower and ***portA*** operate as input ports.
1000110X	***PortC*** upper and ***portA*** operate as output ports while ***portC*** lower and ***portB*** operate as input ports.
1000111X	***PortC*** upper operate as output ports while ***port***C lower, ***portB***, and ***portA*** operates as input ports.
1001000X	***PortC*** lower, ***portB***, and ***portA*** operate as output ports while ***portC*** upper operates as an input port.
1001001X	***PortC*** lower and ***portB*** operate as output ports while ***portC*** upper and **portA** operate as input ports.
1001010X	***PortC*** lower and ***portA*** operate as output ports while ***portC*** upper and ***portB*** operate as input ports.
1001011X	***PortC*** lower operates as an output port while ***portC*** upper, ***portB***, and ***portA*** operate as input ports.
1001100X	***PortB*** and ***portA*** operate as output ports while ***portC*** upper and ***portC*** lower operate as input ports.
1001101X	***PortB*** operates as an output port while ***portC*** upper, ***portC*** lower, and ***portA*** operate as input ports.
1001110X	***PortA*** operates as an output port while ***portB***, ***portC*** upper, and ***portC*** lower operate as output ports.
1001111X	***PortA***, ***portB***, ***portC*** upper, and ***portC*** lower operate as input ports.

combinations of functionality for ***portA***, ***portB***, and ***portC*** in mode 0 operation based on the values of ***CWR*** register.

Referring to Table 6.4, bit 7 of ***CWR*** register must be a logic "1" and bit 0 of ***CWR*** is don't care.

TABLE 6.5. Table showing the combinations of functionality of *portA* and *portB* in mode 1 operation

CWR Register Values	Description
101XX00X	***PortB*** and ***PortA*** operate as strobed output ports.
101XX01X	***PortB*** operates as a strobed output port while ***portA*** operates as a strobed input port.
101XX10X	***PortB*** operates as a strobed input port while ***portA*** operates as a strobed output port.
101XX11X	***PortB*** and ***portA*** operate as strobed input ports.

6.2.2 Mode 1 Operation

In mode 1 operation, the PLB can be configured using the ***CWR*** register to have ***portA*** as either a strobed input or strobed output port and ***portB*** as either a strobed input or strobed output port. In mode 1 operation, ***portC*** functions as control ports for ***portA*** and ***portB***. Table 6.5 shows the combinations of functionality for ***portA*** and ***portB*** in mode 1 operation based on the values of ***CWR*** register.

Figure 6.4 shows the interface signals for PLB in mode 1 operation with ***portA*** and ***portB*** as strobed input ports.

Figure 6.5 shows the interface signals for PLB in mode 1 operation with ***portA*** as a strobed input port and ***portB*** as a strobed output port.

Figure 6.6 shows the interface signals for PLB in mode 1 operation with ***portA*** as a strobed output port and ***portB*** as a strobed input port.

Figure 6.7 shows the interface signals for PLB in mode 1 operation with ***portA*** and ***portB*** as strobed output ports.

Referring to Figures 6.4 through 6.7, when ***portA*** operates as a strobed output port, the control signals of ***portC*** function as follows:

1. ***obfab*** is reset by the rising edge of ***wrb***, set by the falling edge of ***ackab***.
2. ***intra*** is reset by the falling edge or ***wrb***, set by the rising edge of ***ackab***.

Similarly when ***portB*** operates as a strobed output port, the control signals of ***portC*** functions as follows:

1. ***obfbb*** is reset by the rising edge of ***wrb***, set by the falling edge of ***ackbb***.
2. ***intrb*** is reset by the falling edge of ***wrb***, set by the rising edge of ***ackbb***.

If ***portA*** operates as a strobed input port, the control signals of ***portC*** functions as follows:

1. ***ibfa*** is set by the falling edge of ***stbab***, reset by the rising edge of ***rdb***.
2. ***intra*** is set by the rising edge of ***stbab***, reset by the falling edge of ***rdb***.

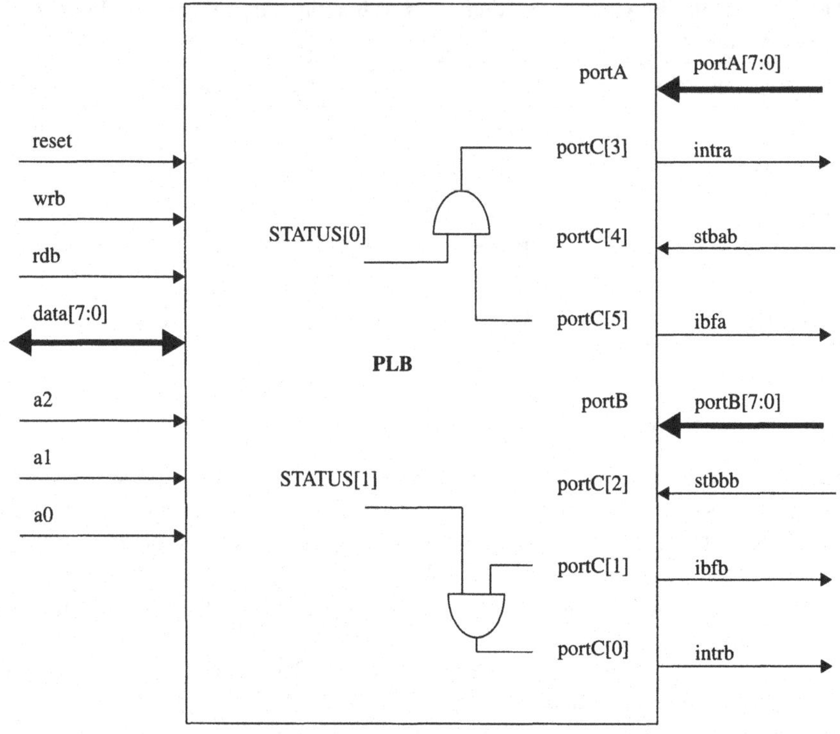

CWR register-101XX11X

FIGURE 6.4. Diagram showing interface signals in mode 1 operation with ***portA*** and ***portB*** as strobed input port.

Similarly when ***portB*** operates as a strobed input port, the control signals of ***portC*** functions as follows:

1. ***ibfb*** is set by the falling edge of ***stbbb***, reset by the rising edge of ***rdb***.
2. ***intrb*** is set by the rising edge of ***stbbb***, reset by the falling edge of ***rdb***.

6.2.3 Mode 2 Operation

In mode 2 operation, ***portA*** operates as a strobed I/O port while ***portB*** can be configured using the ***CWR*** register to be either an input or output port (operation of ***portB*** in mode 2 is the same as operation of ***portB*** in mode 0). In mode 2 operation, ***portC*** functions as control ports for ***portA***. Table 6.6 shows the combinations of functionality of ***portA*** and ***portB*** in mode 2 operation based on the values of ***CWR*** register.

Figure 6.8 shows the interface signals for PLB in mode 2 operation with ***portA*** as a strobed I/O port and ***portB*** as an output port.

Figure 6.9 shows the interface signals for PLB in mode 2 operation with ***portA*** as a strobed I/O port and ***portB*** as an input port.

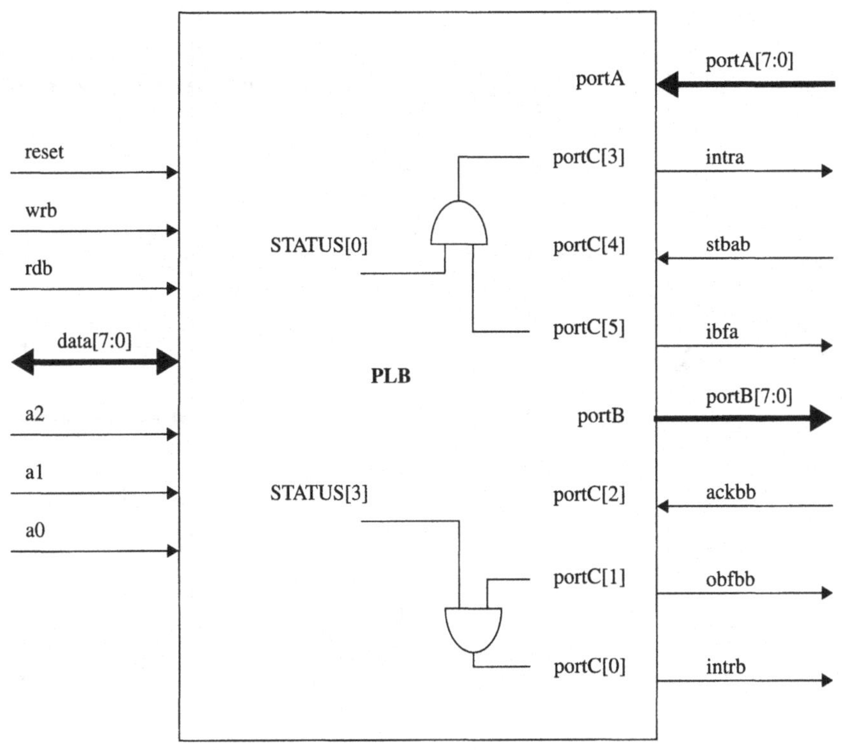

CWR register-101XX01X

FIGURE 6.5. Diagram showing interface signals in mode 1 operation with ***portA*** as strobed input port and ***portB*** as strobed output port.

TABLE 6.6. The combinations of functionality of *portA* and *portB* in mode 1 operation

CWR Register Values	Description
110XX0XX	***PortB*** operates as an output port and ***PortA*** operates as a strobed I/O port.
101XX1XX	***PortB*** operates as an input port while ***portA*** operates as a strobed I/O port.

6.3 MICRO-ARCHITECTURE DEFINITION FOR PROGRAMMABLE PERIPHERAL INTERFACE

Based on the design specification from Section 6.2, a microarchitecture can be derived for the design. The requirements of a ***CWR*** register would point to the need of an eight-bit register, with each bit being able to control the mode

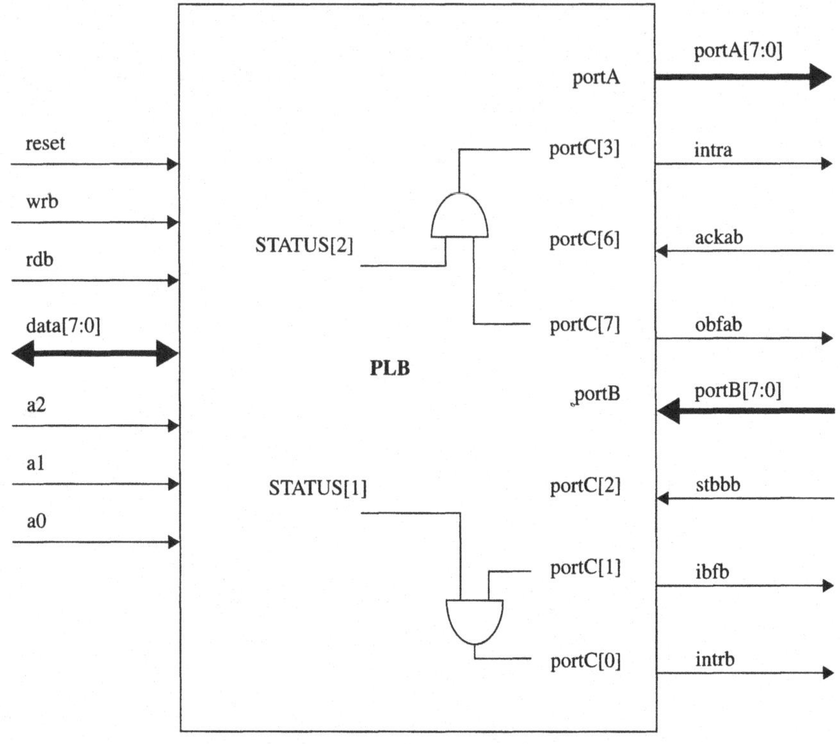

CWR register-101XX10X

FIGURE 6.6. Diagram showing interface signals in mode 1 operation with ***portA*** as strobed output port and ***portB*** as strobed input port.

of operation as well as the functionality of the PLB ports. Similarly, the requirement of a ***STATUS*** register would also point to the need of another eight-bit register with its content being able to control certain control ports of ***portC*** during mode 1 and mode 2 of operation of the PLB. Figure 6.10 shows a block-level diagram of the microarchitecture of the PLB.

Referring to Figure 6.10, a decoder is used to decode the address (***a2, a1, a0***) and control pins (***wrb, rdb, reset***) to determine if the write/read command being issued refers to the ***CWR*** register, ***STATUS*** register, ***portA***, ***portB***, or ***portC***. The decoder also decodes the necessary signals to determine the mode of operation and the functionality of each port (whether its an input, output, strobed input, strobed output, or strobed I/O).

Four sets of flops are used to latch in the values of ***portA***, ***portB***, ***portC***, and ***data***. These flops values from ***portA***, ***portB***, and ***portC*** are multiplexed to the ***data*** port, while the flop values from the data port are demultiplexed to either ***portA***, ***portB***, or ***portC***. (The synthesized Verilog code shown in

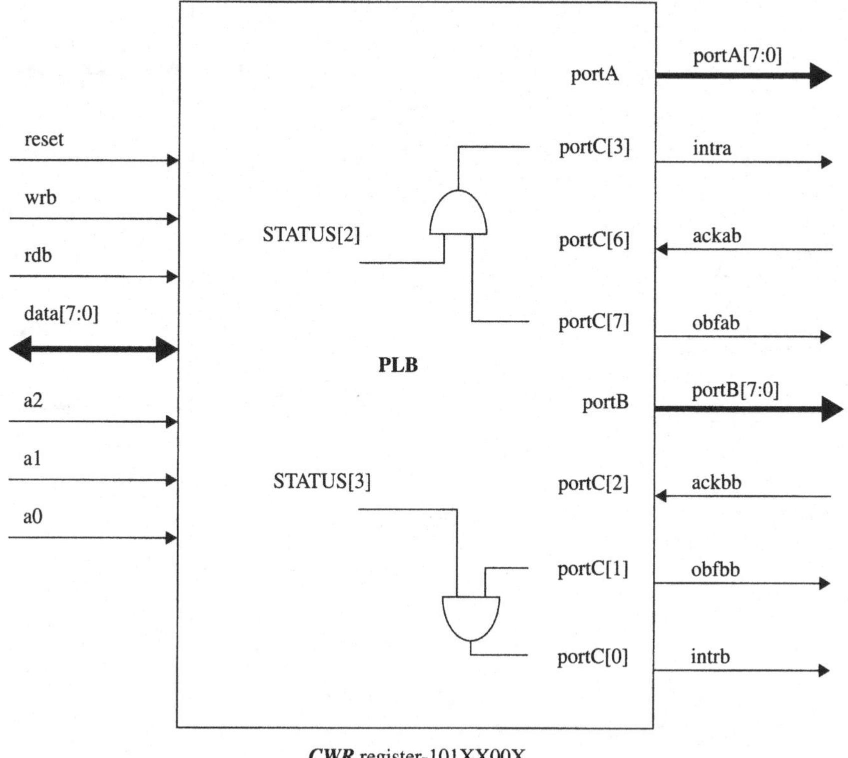

CWR register-101XX00X

FIGURE 6.7. Diagram showing interface signals in mode 1 operation with ***portA*** and ***portB*** as strobed output port.

Example 6.1 uses two additional sets of flop, one each for ***portA*** and ***portB***. These additional flops are to latch in values at ***portA*** and ***portB*** for mode 1 and mode 2 operation when ***portA*** and ***portB*** operates in strobed input, strobed output, or strobed I/O.)

6.4 FLOW DIAGRAM DEFINITION FOR PROGRAMMABLE PERIPHERAL INTERFACE

Before a synthesizable Verilog code is written for the programmable peripheral interface, a designer should always spend some time drawing a flow diagram to represent the process flow of the design. Such a diagram can be very useful when the designer starts to write the Verilog code for the design.

The designer does not need to restrict himself/herself with only using flow diagrams. The designer can choose to draw state diagrams, bubble

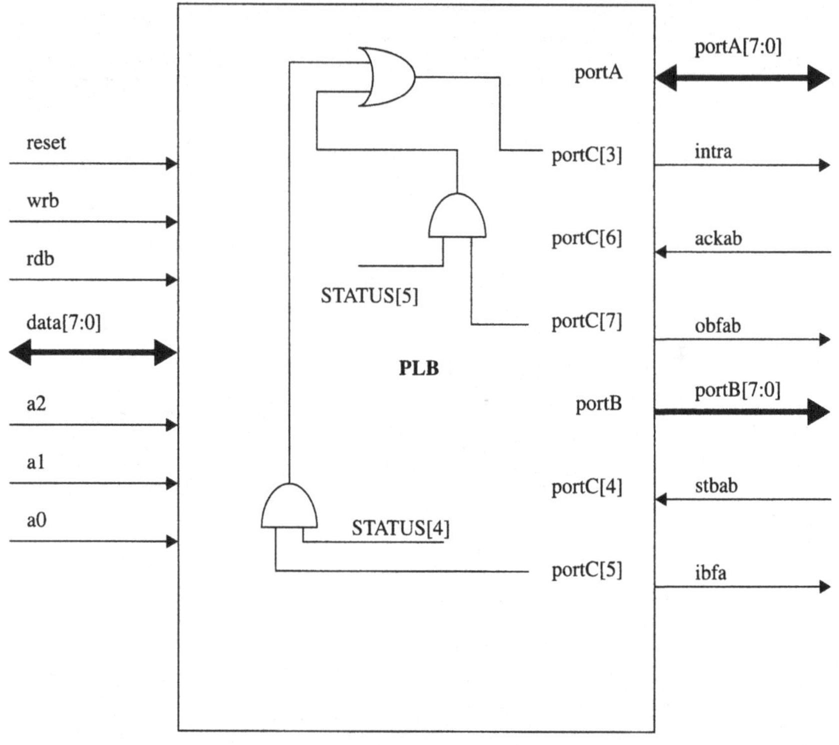

CWR register-101XX0XX

FIGURE 6.8. Interface signals in mode 2 operation with ***portA*** as strobed I/O and ***portB*** as output port.

diagrams, or even pseudocode, depending on which is more suitable for the designer.

Figure 6.11 shows a flow diagram that writes data to the ***CWR*** register and ***STATUS*** register. It covers the condition that the user can use to write data into the ***CWR*** register or the ***STATUS*** register.

Referring to Figure 6.11:

1. If rising edge of ***reset*** is detected, the ***STATUS*** register is reset to a value of "00000000" and the ***CWR*** register is reset to "10011110" (which defaults to the PLB to operate in mode 0 with ***portC***, ***portB*** and ***portA*** as input ports).
2. If rising edge or ***wrb*** is detected, address is checked for "011" or "111". If address is "011", the data at data bus is written into the ***CWR*** register, and if address is "111", the data at data bus is written into the ***STATUS*** register.

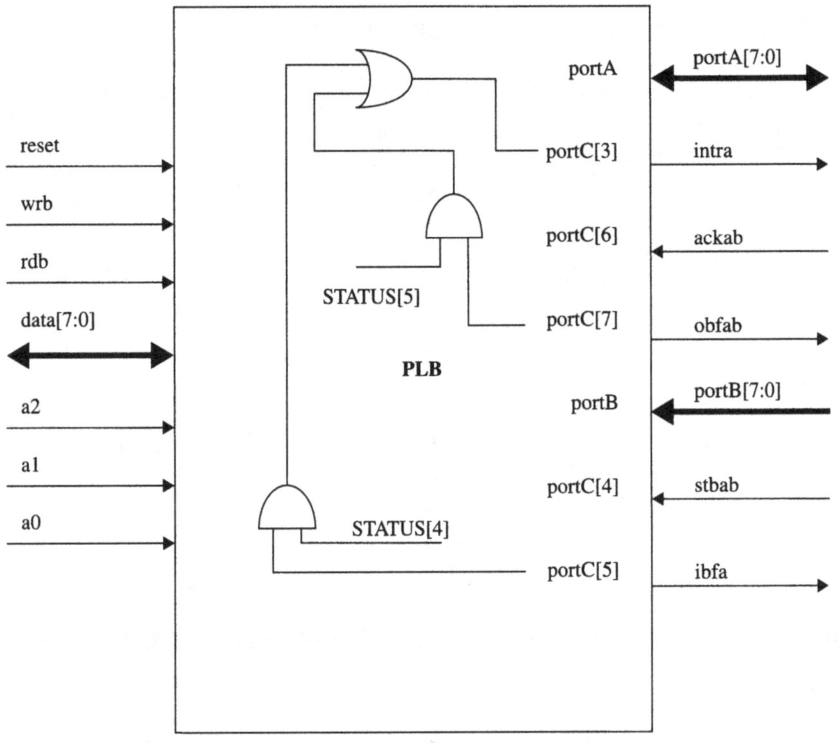

CWR register-101XX1XX

FIGURE 6.9. Interface signals in mode 2 operation with ***portA*** as strobed I/O and ***portB*** as input port.

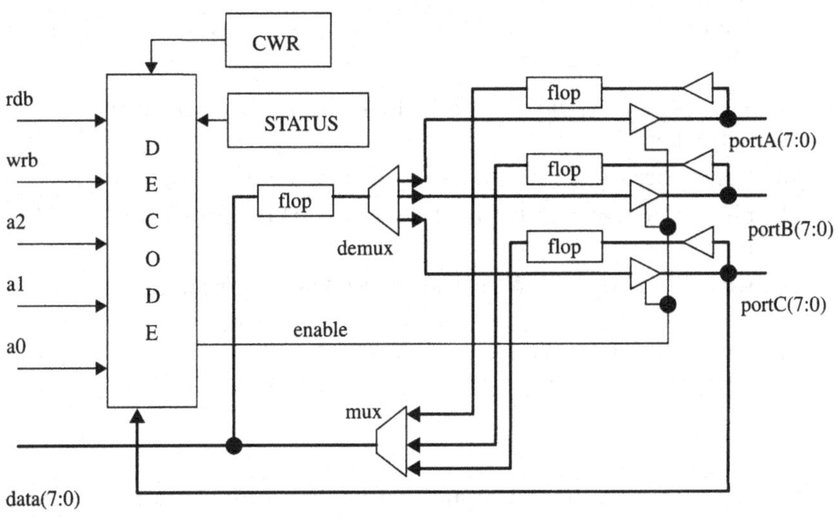

FIGURE 6.10. Microarchitectural block diagram of the programmable peripheral interface design.

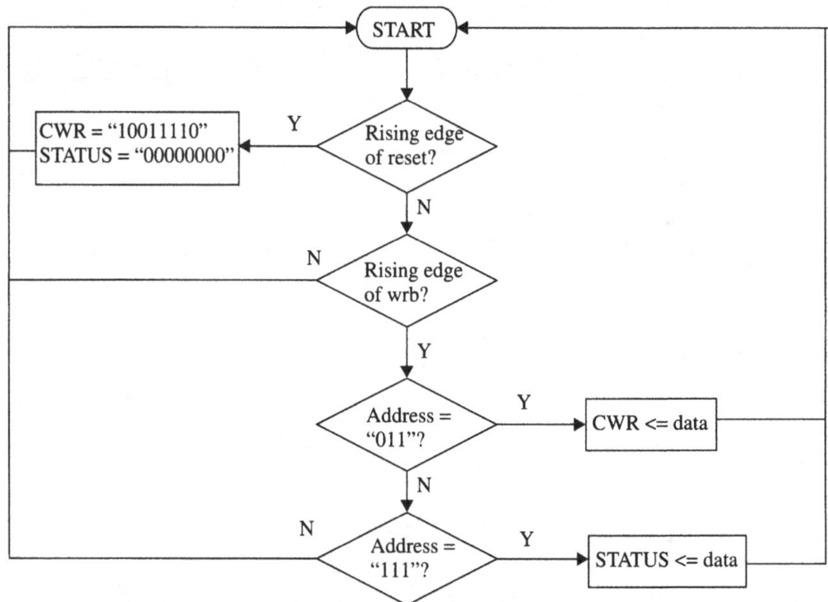

FIGURE 6.11. Flow diagram for writing data to ***CWR*** register and ***STATUS*** register.

Figure 6.12 shows a flow diagram that reads the contents of CWR register and ***STATUS*** register. Referring to Figure 6.12:

1. If ***int_reset*** is a logic high, ***data*** bus is tri-stated. Signal ***int_reset*** is a logic condition that is derived from the BOOLEAN expression:

   ```
   int_reset = reset | ~CWR[7] | (~wrb & ((address =
   ‘011’) | (address = “111”)))
   ```

   This would ensure ***int_reset*** is at logic "1" (causing ***data*** bus to be tri-stated) when:

   a. a reset occurs,

   b. bit 7 of ***CWR*** register is at logic "0". This means that the PLB is inactive, or

   c. when a write to ***CWR*** register or ***STATUS*** register occurs.

2. If ***rdb*** is detected as a logic low, address is checked for "011" or "111". If address is "011", the contents of ***CWR*** register are read to the data bus. If address is "111", the contents of ***STATUS*** register are read to the ***data*** bus.

Figure 6.13 shows a flow diagram that latches in the data from ***data*** bus onto ***latch_data*** bus. Referring to Figure 6.13:

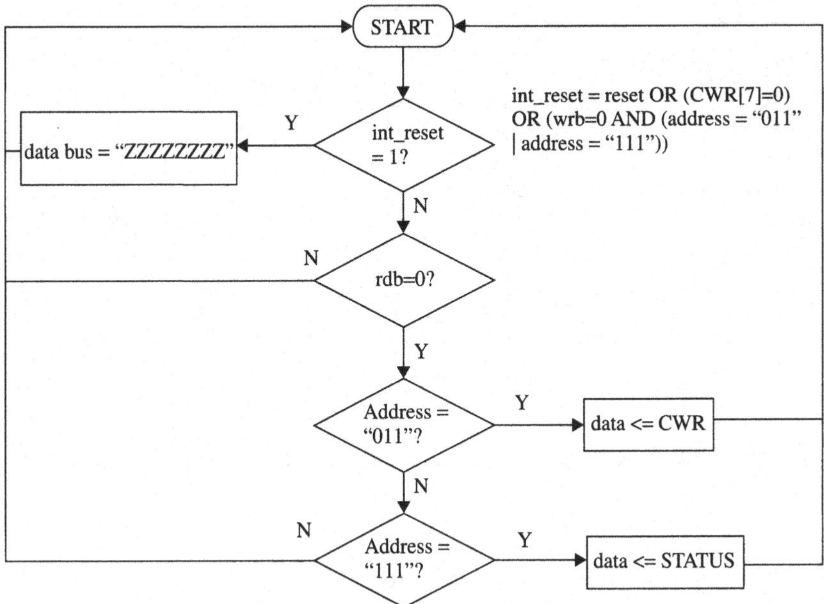

FIGURE 6.12. Flow diagram for reading contents of ***CWR*** register and ***STATUS*** register.

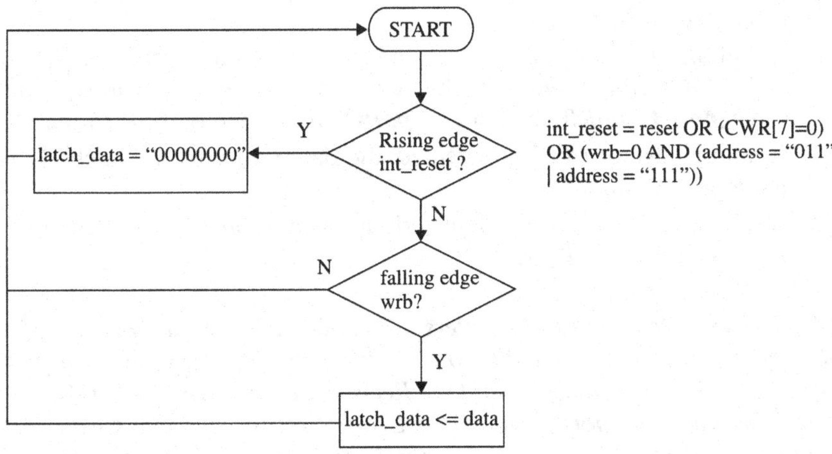

FIGURE 6.13. Flow diagram for Latching Data from ***data*** bus onto ***latch_data*** bus.

1. If ***int_reset*** is a logic high, ***latch_data*** bus is reset to "00000000".
2. if falling edge of ***wrb*** is detected, the data on ***data*** bus are latched onto ***latch_data*** bus.

Figure 6.14 shows a flow diagram that latches in the data from ***portA***, ***portB***, and ***portC*** onto different buses. Referring to Figure 6.14:

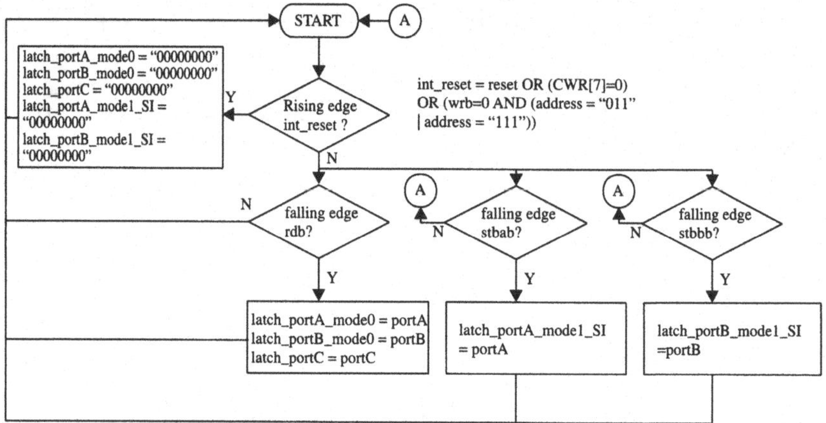

FIGURE 6.14. Flow diagram for Latching Data from ***portA***, ***portB***, and ***portC***.

1. If rising edge of ***int_reset*** is detected,

   latch_portA_mode0, latch_portB_mode0, latch_portA_mode1_SI, latch_portB_mode1_SI, latch_portC

   is reset to "00000000"
2. If falling edge of ***rdb*** is detected, the data on ***portA*** are latched onto ***latch_portA_mode0***, the data on ***portB*** are latched onto ***latch_portB_mode0***, the data on ***portC*** are latched onto ***latch_portC***.
3. If falling edge of ***stbab*** is detected, the data on ***portA*** are latched onto ***latch_portA_mode1_SI***.
4. If falling edge of ***stbbb*** is detected, the data on ***portB*** are latched onto ***latch_portB_mode1_SI***.

Note: There are two sets of registers to latch in the values at ***portA***. *Register* ***latch_portA_mode0***, *as its name implies, is for latching in data from* ***portA*** *when the PLB is operating in mode 0. Register* ***latch_portA_mode1_SI*** *is for latching in data from* ***portA*** *when the PLB is operating in mode 1 with* ***portA*** *as a strobed input port. Similarly for* ***portB***, *there are also two sets of registers for latching data. There are, however, only one set of registers for latching in data from* ***portC*** *because* ***portC*** *can only operate as an input port in mode 0 (in mode 1 and mode 2,* ***portC*** *operates as control signals).*

Figure 6.14 shows the flow diagram for latching in data from ***portA***, ***portB***, and ***portC*** into registers internally in the PLB. The data in these registers are read from the PLB during a read condition. Figure 6.15 shows the flow diagram for reading data from these registers to external devices using ***data*** bus.

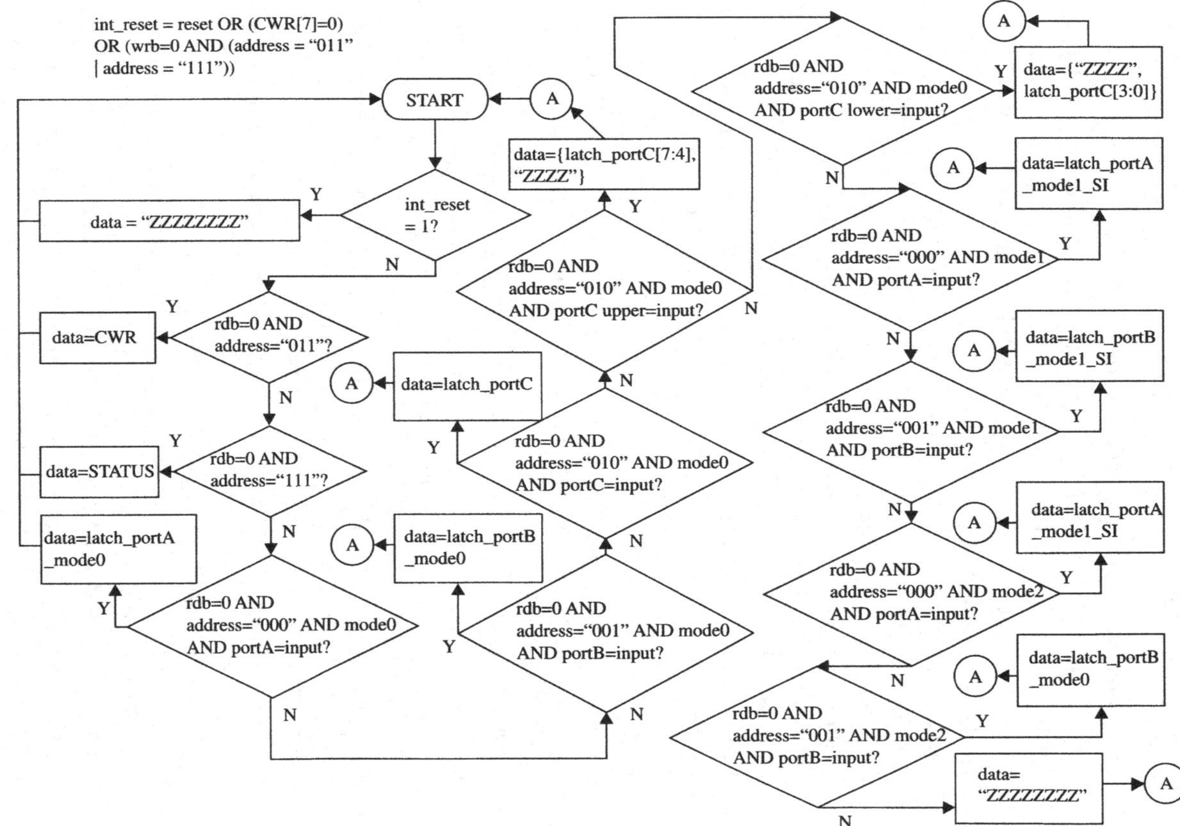

FIGURE 6.15. Flow diagram for latching Ddata from ***portA***, ***portB***, and ***portC***.

Referring to Figure 6.15:

1. If ***int_reset*** is at logic "1", ***data*** bus is tri-stated.
2. If ***rdb*** is at logic "0" and address is at "011", the contents of ***CWR*** register are read onto the ***data*** bus.
3. If ***rdb*** is at logic "0" and address is at "111", the contents of ***STATUS*** register are read onto the ***data*** bus.
4. If ***rdb*** is at logic "0", address is at "000", PLB is operating in mode 0, and ***portA*** is operating as an input port, the contents of ***latch_portA_mode0*** register are read onto the ***data*** bus.
5. If ***rdb*** is at logic "0", address is at "001", PLB is operating in mode 0, and ***portB*** is operating as an input port, the contents of ***latch_portB_mode0*** register are read onto the ***data*** bus.
6. If ***rdb*** is at logic "0", address is at "010", PLB is operating in mode 0, and ***portC*** is operating as input, the contents of ***latch_portC*** register are read onto the ***data*** bus.
7. If ***rdb*** is at logic "0", address is at "010", PLB is operating in mode 0, and ***portC*** upper (***portC[7:4]***) is operating as input (***portC*** lower is operating as output), the contents of ***latch_portC[7:4]*** register are read onto the ***data*** bus.
8. If ***rdb*** is at logic "0", address is at "010", PLB is operating in mode 0, and ***portC*** lower (***portC[3:0]***) is operating as input (***portC*** upper is operating as output), the contents of ***latch_portC[3:0]*** register are read onto the ***data*** bus.
9. If ***rdb*** is at logic "0", address is at "000", PLB is operating in mode 1, and ***portA*** is operating as strobed input, the contents of ***latch_portA_mode1_SI*** register are read onto the ***data*** bus.
10. If ***rdb*** is at logic "0", address is at "001", PLB is operating in mode 1, and ***portB*** is operating as strobed input, the contents of ***latch_portB_mode1_SI*** register are read onto the ***data*** bus.
11. If ***rdb*** is at logic "0", address is at "000", PLB is operating in mode 2, and ***portA*** is operating as strobed I/O, the contents of ***latch_portA_mode1_SI*** register are read onto the ***data*** bus.
12. If ***rdb*** is at logic "0", address is at "001", PLB is operating in mode 2, and ***portB*** is operating as input (in mode 2, operation of ***portB*** is the same as in mode 0), the contents of ***latch_portB_mode0*** register are read onto the ***data*** bus.
13. If none of the conditions specified in (1) to (12) are met, the ***data*** bus is tri-stated.

Figure 6.16 shows the flow diagram for generating the enable signals that determine whether ***portA***, ***portB***, and ***portC*** are input ports or output ports in mode 0.

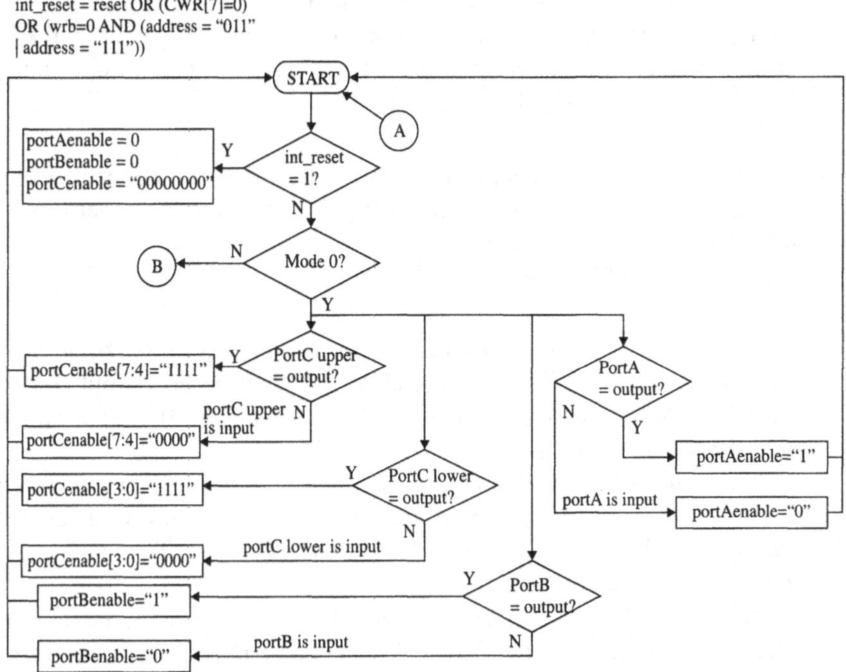

FIGURE 6.16. Flow diagram for generating the enable signals for ***portA***, ***portB***, and ***portC*** in mode 0.

Referring to Figure 6.16:

1. If ***int_reset*** is at logic "1", ***portAenable***, ***portBenable***, and ***portCenable*** are driven to logic "0".
2. If PLB is in mode 0 and ***CWR***[4] is at logic "0", the upper bits of ***portCenable*** (bits 7 to 4) are driven to logic "1". This would enable the upper bits of ***portC*** to be an output port. If ***CWR***[4] is at logic "1", the upper bits of ***portCenable*** (bits 7 to 4) are driven to logic "0". This would enable the upper bits of ***portC*** to be an input port.
3. If PLB is in mode 0 and ***CWR***[3] is at logic "0", the lower bits of ***portCenable*** (bits 3 to 0) are driven to logic "1". This would enable the lower bits of ***portC*** to be an output port. If ***CWR***[3] is at logic "1", the lower bits of ***portCenable*** (bits 3 to 0) are driven to logic "0". This would enable the lower bits of ***portC*** to be an input port.
4. If PLB is in mode 0 and ***CWR***[2] is at logic "0", ***portBenable*** is driven to logic "1". This would enable ***portB*** to be an output port. If ***CWR***[2] is at logic "1", ***portBenable*** is driven to logic "0". This would enable ***portB*** to be an input port.

5. If PLB is in mode 0 and ***CWR***[1] is at logic "0", ***portAenable*** is driven to logic "1". This would enable ***portA*** to be an output port. If ***CWR***[1] is at logic "1", ***portAenable*** is driven to logic "0". This would enable ***portA*** to be an input port.

Referring to Figure 6.16, connector B of the flow diagram connects to the flow diagram of Figure 6.17, which shows the flow diagram for generation of enable signals in mode 1 operation.

Referring to Figure 6.17:

1. If PLB is in mode 1 and ***portB*** is a strobed output port, ***portBenable*** is driven to logic "1". This would enable ***portB*** to be an output port. If ***portB*** is strobed input, ***portBenable*** is driven to logic "0", meaning that ***portB*** is an input port.
2. If PLB is in mode 1 and ***portA*** is a strobed output port, ***portAenable*** is driven to logic "1". This would enable ***portA*** to be an output port. If ***portA*** is strobed input, ***portAenable*** is driven to logic "0", meaning that ***portA*** is an input port.
3. If PLB is in mode 1 and ***portA*** and ***portB*** are strobed output ports, ***port-Cenable*** is driven to "10001011". This enables ***portC*** bits 7 (***obfab***), 3 (***intra***), 1 (***obfbb***), and 0(***intrb***) to be an output port, while ***portC*** bit 6(***ackab***), 5 (not used but default to input), 4 (not used but default to input), and 2 (***ackbb***) are input ports.
4. If PLB is in mode 1 and ***portA*** is a strobed input port and ***portB*** is a strobed output port, ***portCenable*** is driven to "00101011". This enables ***portC*** bits 5 (***ibfa***), 3 (***intra***), 1 (***obfbb***), and 0 (***intrb***) to be output ports while ***portC*** bit 7 (not used but default to input), 6 (not used but default to input), 4 (***stbab***), and 2 (***ackbb***) are input ports.
5. If PLB is in mode 1 and ***portA*** is a strobed output port and ***portB*** is a strobed input port, ***portCenable*** is driven to "10001011". This enables ***portC*** bits 7 (***obfab***), 3 (***intra***), 1 (***ibfb***), and 0 (***intrb***) to be output ports while ***portC*** bits 6 (***ackab***), 5 (not used but default to input), 4 (not used but default to input), and 2 (***stbbb***) are input ports.
6. If PLB is in mode 1 and ***portA*** and ***portB*** are strobed input ports, ***portCenable*** is driven to "00101011". This enables ***portC*** bits 5 (***ibfa***), 3 (***intra***), 1 (***ibfb***), and 0 (***intrb***) to be output ports while ***portC*** bits 7 (not used but default to input), 6 (not used but default to input), 4 (***stbab***), and 2 (***stbbb***) are input ports.

Referring to Figure 6.17, connector C of the flow diagram connects to the flow diagram of Figure 6.18, which shows the flow diagram for generating the enable signals in mode 2 operation.

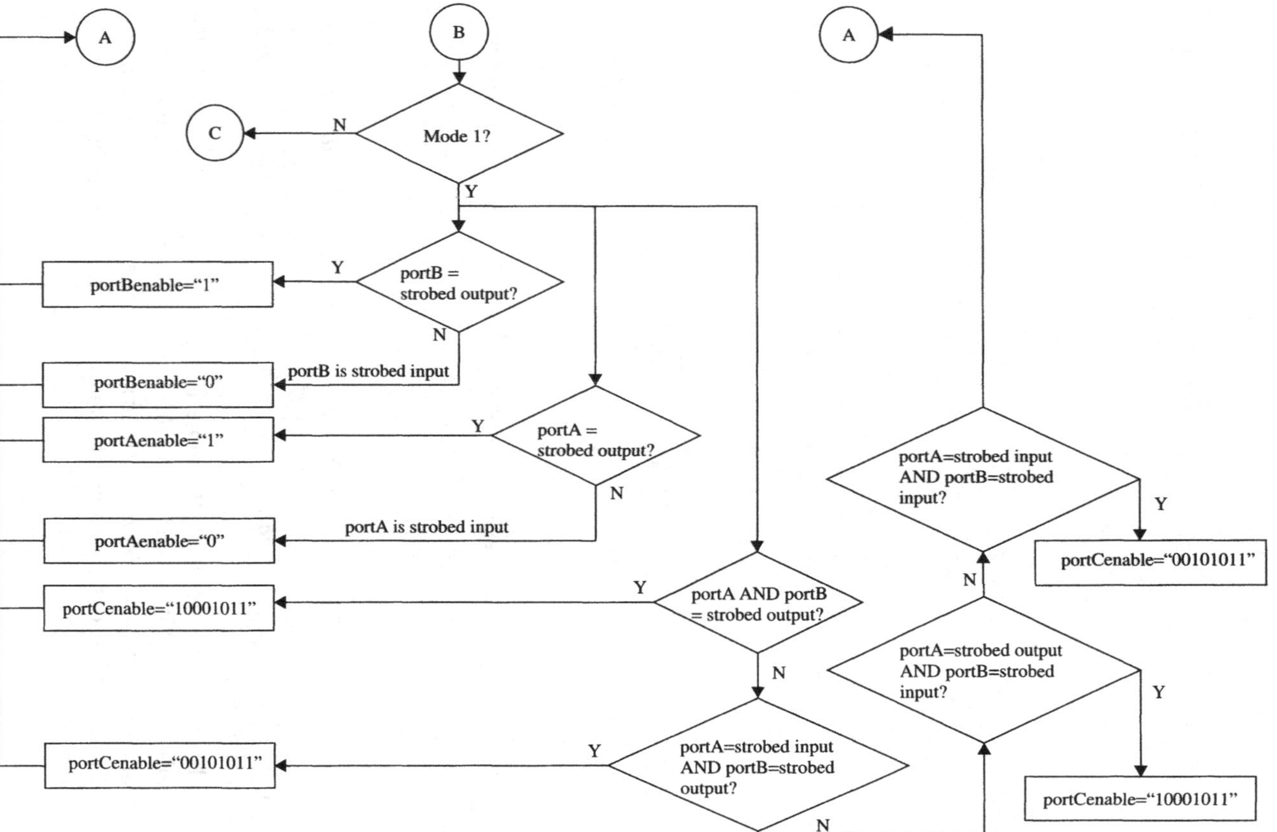

FIGURE 6.17. Flow diagram for generating the enable signals for ***portA***, ***portB***, and ***portC*** in mode 1.

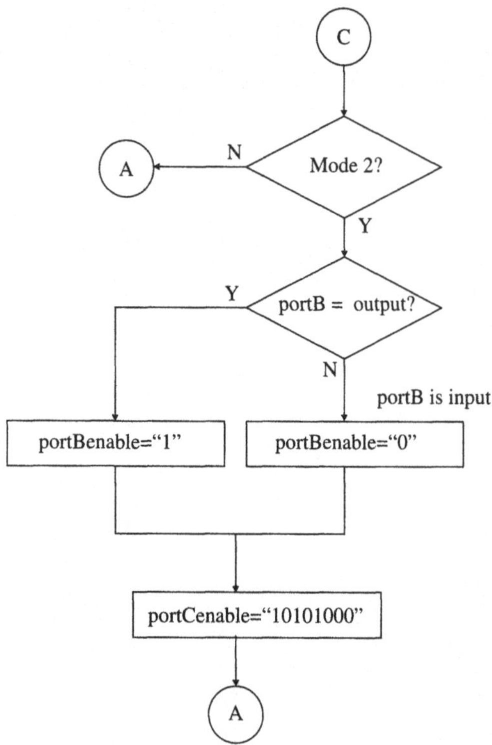

FIGURE 6.18. Flow diagram to generate the enable signals for ***portA***, ***portB***, and ***portC*** in mode 2.

In Figure 6.18:

1. If PLB is in mode 2 and ***portB*** is an output port, ***portBenable*** is driven to logic "1". This would enable ***portB*** to be an output port. If ***portB*** is an input port, ***portBenable*** is driven to logic "0", meaning that ***portB*** is an input port.
2. If PLB is in mode 2, ***portCenable*** is driven to logic "10101000". This enables ***portC*** bits 7 (***obfab***), 5 (***ibfa***), and 3 (***intra***) to be output ports while ***portC*** bits 6 (***ackab***), 4 (***stbab***), 2 (not used but default to input), 1 (not used but default to input), and 0 (not used but default to input) are input ports.

Figure 6.19 shows the flow diagram to generate ***set_obfab*** and ***set_obfbb*** signals that are used to generate ***out_obfab*** and ***out_obfbb*** signals. The ***out_obfab*** and ***out_obfbb*** signals goes directly to the output ***portC[7]*** (***out_obfab***) if ***portA*** is functioning as a strobed output port in mode 1 or strobed I/O port in mode 2, and ***portC[1]*** (***out_obfbb***) if ***portB*** is functioning as a strobed output port in mode 1.

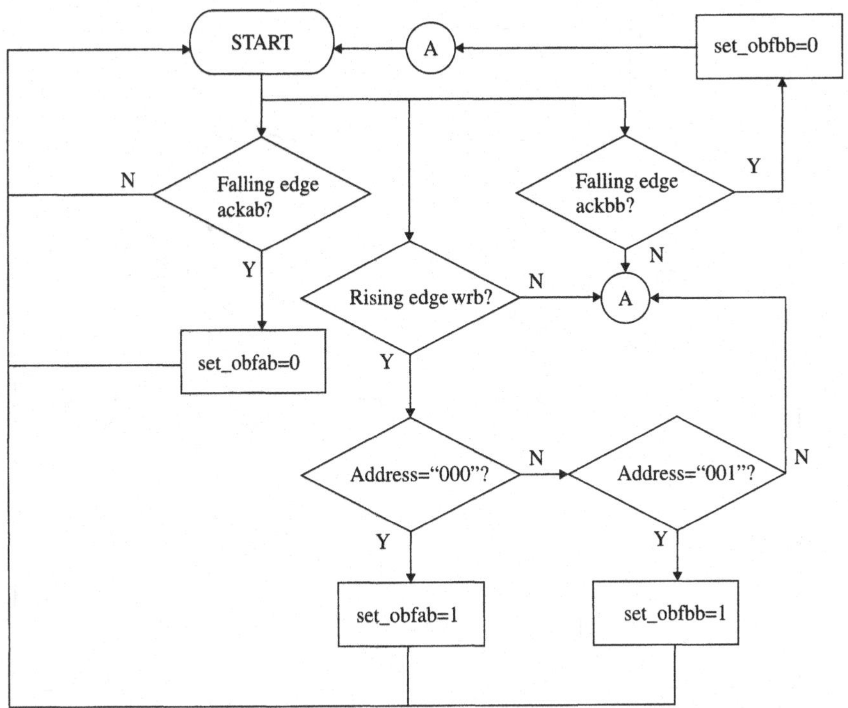

FIGURE 6.19. Diagram showing generation of ***set_obfab*** and ***set_obfbb*** logic.

Referring to Figure 6.19:

1. The falling edge of ***ackab*** resets ***set_obfab*** to logic "0".
2. The falling edge of ***ackbb*** resets ***set_obfbb*** to logic "0".
3. If a rising edge of ***wrb*** is detected, the address is checked for a logic value of "000" or "001". If the address is at logic value "000", ***set_obfab*** is set to logic "1", and if address is at logic value "001", ***set_obfbb*** is set to logic "1".

Figure 6.20 shows the flow diagram to generate ***out_obfab*** and ***out_obfbb*** signals for mode 1 and mode 2 operations. In Figure 6.20:

- For generation of ***out_obfab***:
  1. If the PLB is operating in mode 1, ***portA*** is an output port, the address refers to neither ***CWR*** nor ***STATUS*** registers, and ***set_obfab*** is at logical "1", ***out_obfab*** is driven to logic "0".
  2. If the PLB is operating in mode 2, the address refers to neither ***CWR*** nor ***STATUS*** registers, and ***set_obfab*** is at logical "1", ***out_obfab*** is driven to logic "0".
  3. Otherwise, ***out_obfab*** is driven to logic "1".

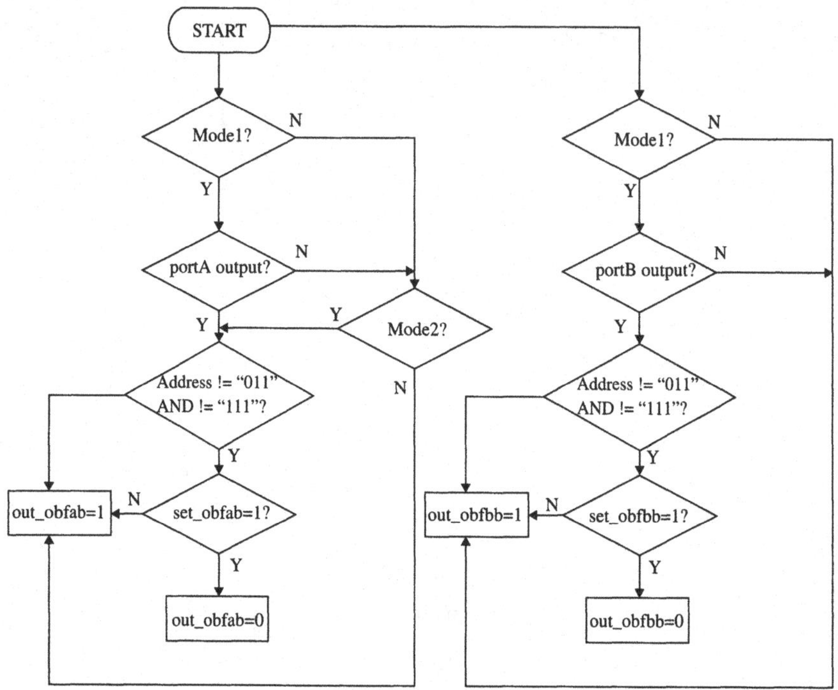

FIGURE 6.20. Diagram showing generation of ***out_obfab*** and ***out_obfbb*** logic.

- For generation of ***out_obfbb***:
  1. If the PLB is operating in mode 1, ***portB*** is an output port, the address refers to neither ***CWR*** nor ***STATUS*** registers, and ***set_obfbb*** is at logical "1", ***out_obfbb*** is driven to logic "0".
  2. Otherwise, ***out_obfbb*** is driven to logic "1". The flow diagram for generation of ***out_obfbb*** does not need to check for the condition of mode 2 because, in mode 2 of operation, ***portB*** functions identically as it does in mode 0 (which means it can only function as an input port or output port, and not a strobed input or strobed output port).

Figure 6.21 shows the flow diagram to generate ***ackab*** and ***ackbb*** signals for mode 1 and mode 2 operations. In Figure 6.21:

- For generation of ***ackab***:
  1. If PLB is operating in mode 1 and ***portA*** functions as a strobed output port, ***ackab*** is driven by ***portC***[6]. If ***portC***[6] is at logic "1", ***ackab*** is at logic "1", and if ***portC***[6] is at logic "0", ***ackab*** is at logic "0".
  2. If PLB is operating in mode 2, ***ackab*** is driven by ***portC***[6]. In mode 2, the flow diagram does not check if ***portA*** functions as a strobed

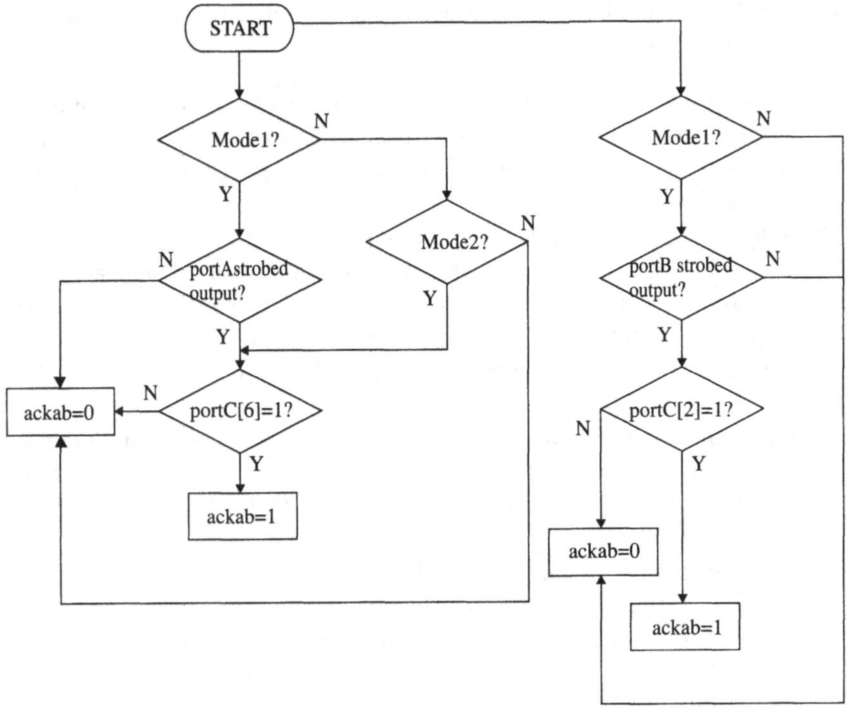

FIGURE 6.21. Diagram showing generation of ***ackab*** and ***ackbb*** logic.

output (or strobed input), because in mode 2, ***portA*** can only function as strobed I/O port.

- For generation of ***ackbb***:
  1. If PLB is operating in mode 1 and ***portB*** functions as a strobed output port, ***ackbb*** is driven by ***portC***[2]. If ***portC***[2] is at logic "1", ***ackbb*** is at logic "1", and if ***portC***[2] is at logic "0", ***ackbb*** is at logic "0".
  2. The flow diagram for generation of ***ackbb*** does not need to check for the condition of mode 2, because in mode 2 of operation, ***portB*** functions identically as it does in mode 0 (which means it can only function as an input or output port, and not a strobed input or output).

For operation of PLB in mode 1 or mode 2, the signals ***intra*** and ***intrb*** need to be generated. These signals are represented by ***portC[3]*** (***intra***) if ***portA*** is operating in mode 1 or in mode 2, and ***portC[0]*** (***intrb***) if ***portB*** is operating in mode 1.

In order to generate the ***intra*** signal for operation of PLB in mode 1 or mode 2 operation, four signals are created internally in the PLB. These four signals (***set_so_intra, wrb_portA, rdb_portA, set_si_intra***) capture a specific condition for which the ***out_intra*** is generated. To generate ***out_intra*** in mode 1 and mode 2 operation:

1. ***set_so_intra.*** This signal is set only when ***portA*** is a strobed output in mode 1 operation or when ***portA*** is a strobed I/O in mode 2 operation. ***set_so_intra*** is reset by the falling edge of ***wrb_portA*** and is set by rising edge ***ackab***.
2. ***wrb_portA.*** This signal follows ***wrb*** but only when the address is showing a ***portA*** access (address = "000"). This signal is used instead of ***wrb*** because, when a falling edge of ***wrb*** is detected, it does not necessary correspond to a ***portA*** access. Therefore, ***wrb_portA*** is created to ensure that the falling edge of ***wrb_portA*** refers to the falling edge of ***wrb*** for ***portA*** access.
3. ***set_si_intra.*** This signal is set only when ***portA*** is a strobed input in mode 1 operation or when ***portA*** is a strobed I/O in mode 2 operation. ***set_si_intra*** is reset by the falling edge of ***rdb_portA*** and is set by the rising edge ***stbab***.
4. ***rdb_portA.*** This signal follows ***rdb*** but only when the address shows a ***portA*** access (address = "000"). This signal is used instead of ***rdb*** because, when a falling edge of ***rdb*** is detected, it does not necessarily correspond to a ***portA*** access. Therefore, ***rdb_portA*** is created to ensure that the falling edge of ***rdb_portA*** refers to the falling edge of ***rdb*** for ***portA*** access.

Figure 6.22 shows the flow diagram for generating ***set_so_intra*** and Figure 6.23 shows the flow diagram for generating ***set_si_intra***.

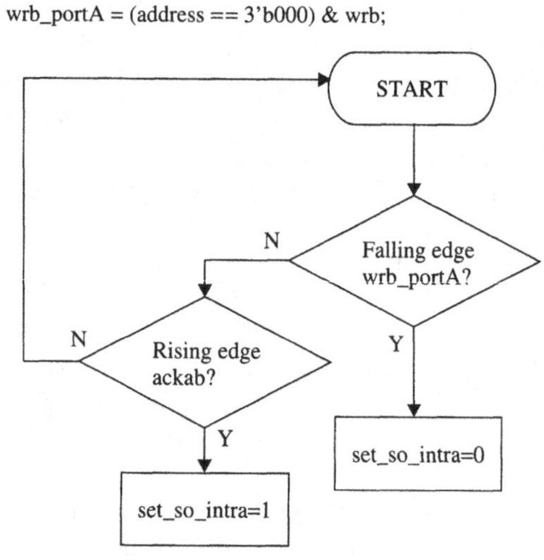

FIGURE 6.22. Diagram showing generation of ***set_so_intra*** logic.

rdb_portA = (address = = 3'b000) & rdb;

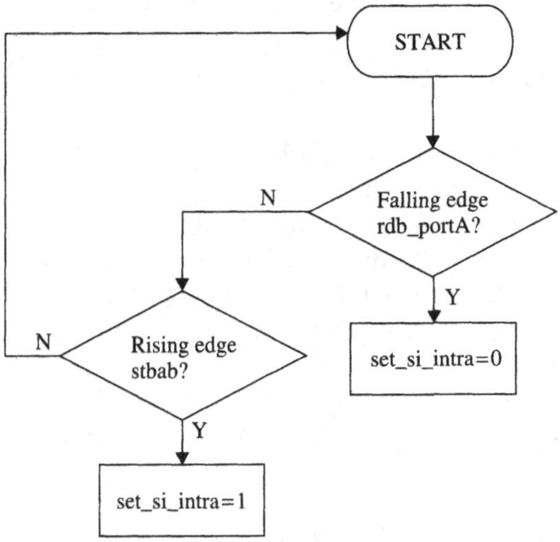

START

Falling edge rdb_portA?

N

Y

set_si_intra=0

Rising edge stbab?

N

Y

set_si_intra=1

FIGURE 6.23. Diagram showing generation of ***set_si_intra*** logic.

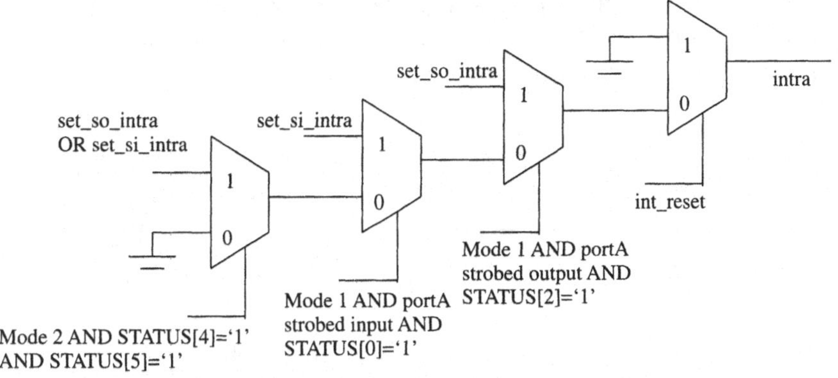

FIGURE 6.24. Diagram showing generation of ***intra*** logic.

Based on the generated signals of ***set_si_intra*** and ***set_so_intra***, ***intra*** can be generated. Figure 6.24 shows four 2-to-1 multiplexers for generation of ***intra***.

Referring to Figure 6.24:

1. If ***int_reset*** is at logic "1", ***intra*** is driven to logic "0".
2. If PLB is operating in mode 1, ***portA*** functions as a strobed output and ***STATUS*** bit 2 is at logic "1", ***intra*** is driven by ***set_so_intra***.
3. If PLB is operating in mode 1, ***portA*** functions as at strobed input and ***STATUS*** bit 0 is a logic "1", ***intra*** is driven by ***set_si_intra***.

4. If PLB is operating in mode 2, ***STATUS*** bit 4 and bit 5 are at logic "1", ***intra*** is driven to logic "1" if either ***set_so_intra*** or ***set_si_intra*** is at logic "1".
5. Otherwise, ***intra*** is driven to logic "0".

Note: The circuit shown in Figure 6.24 uses four 2-to-1 multiplexers. A single 4-to-1 multiplexer can also be used.

The logic required to generate ***intrb*** is similar to the logic required to generate ***intra***. Figure 6.25 shows the flow diagram for generation of ***set_so_intrb*** (similar to Figure 6.22 for generation of ***set_so_intra***). Figure 6.26 shows the flow diagram for generation of ***set_si_intrb*** (similar to Figure 6.23 for generation of ***set_si_intra***).

Based on the generated signals of ***set_si_intrb*** and ***set_so_intrb***, ***intrb*** can be generated. Figure 6.27 shows three 2-to-1 multiplexers for generation of ***intrb***. In Figure 6.27:

1. If ***int_reset*** is at logic "1", ***intrb*** is driven to logic "0".
2. If PLB is operating in mode 1, ***portB*** functions as a strobed output and ***STATUS*** bit 3 is at logic "1", ***intrb*** is driven by ***set_so_intrb***.
3. If PLB is operating in mode 1, ***portB*** functions as a strobed input, and ***STATUS*** bit 1 is at logic "1", ***intrb*** is driven by ***set_si_intrb***.
4. Otherwise, ***intrb*** is driven to logic "0".

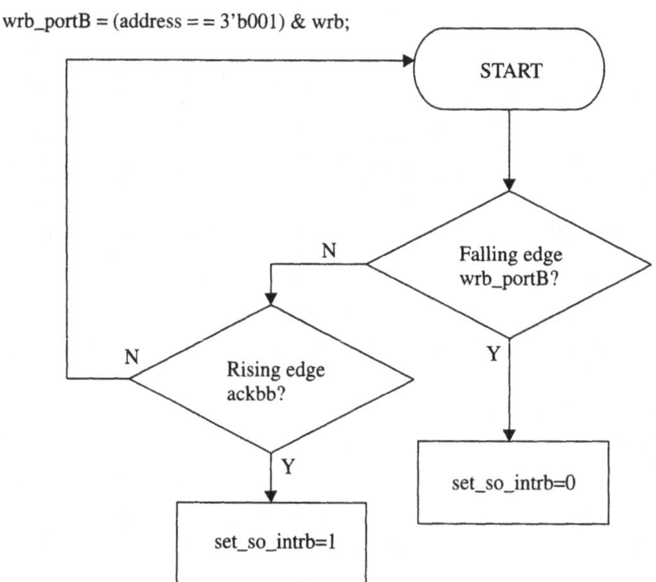

FIGURE 6.25. Diagram showing generation of ***set_so_intrb*** logic.

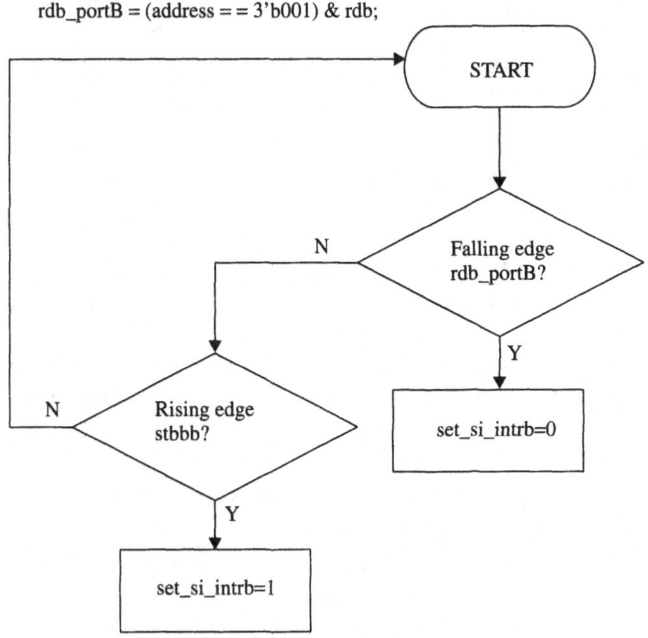

FIGURE 6.26. Diagram showing generation of ***set_si_intrb*** logic.

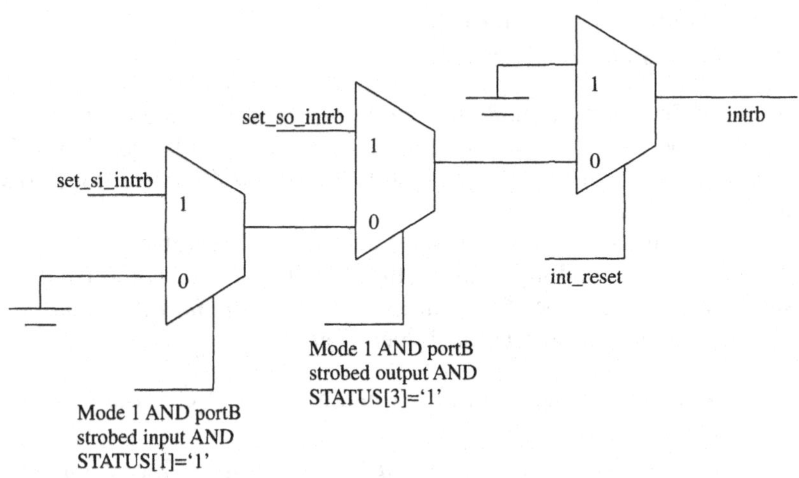

FIGURE 6.27. Diagram showing generation of ***intrb*** logic.

Note: The circuit shown in Figure 6.27 uses three 2-to-1 multiplexers. A single 4-to-1 multiplexer can also be used.

Figure 6.28 shows a flow diagram for generation of ***stbab*** and ***stbbb***. In Figure 6.28, for generation of ***stbab***:

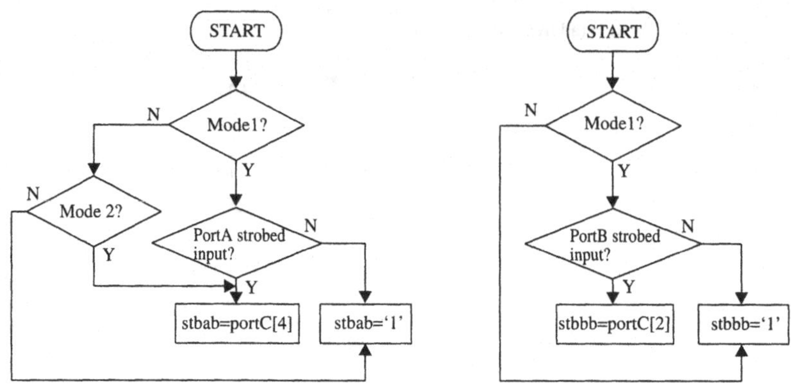

FIGURE 6.28. Diagram showing generation of ***stbab*** and ***stbbb*** logic.

1. If the PLB is operating in mode 1, ***portA*** functions as a strobed input port, ***stbab*** is driven by ***portC***[4].
2. If the PLB is operating in mode 2, ***stbab*** is driven by ***portC***[4].
3. Otherwise, ***stbab*** is driven to logic "1".

Referring to Figure 6.28, for generation of ***stbbb***:

1. If the PLB is operating in mode 1, ***portB*** functions as a strobed input port, ***stbbb*** is driven by ***portC***[2].
2. Otherwise, ***stbbb*** is driven to logic "1".

For operation of PLB in mode 1 or mode 2 of operation, the signal ***ibfa*** and ***ibfb*** needs to be generated. These signals are represented by ***portC[5]*** (***ibfa***) if ***portA*** is operating in mode 1 or in mode 2, and ***portC[1]*** (***ibfb***) if ***portB*** is operating in mode 1.

In order to generate the ***ibfa*** signal for operation of PLB in mode 1 or mode 2, two signals are created internally in the PLB. These two signals (***reset_ibfa*** and ***set_ibfa***) capture a specific condition for which the ***ibfa*** is generated. To generate ***ibfa*** in mode 1 and mode 2 operation:

1. ***reset_ibfa.*** This signal is set only when a rising edge of ***rdb*** is detected and ***set_ibfa*** is at a logic "1".
2. ***set_ibfa.*** This signal is reset when ***int_reset*** or ***reset_ibfa*** is at a logic "1". It is set when PLB is operating in mode 1, ***portA*** functions as a strobed input port, ***reset_ibfa*** is at logic "0", and ***stbab*** is at logic "0". This signal is also set when PLB is operating in mode 2, ***reset_ibfa*** is at logic "0", and ***stbab*** is at logic "0".

Figure 6.29 shows the flow diagram for generation of ***set_ibfa*** and ***reset_ibfa***. In Figure 6.29:

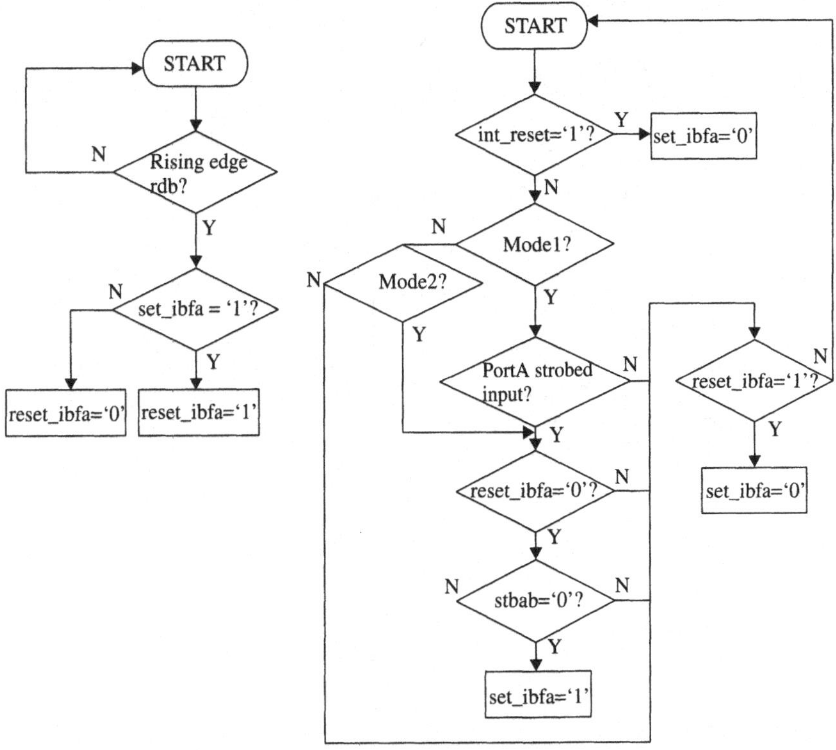

FIGURE 6.29. Diagram showing generation of ***set_ibfa*** and ***reset_ibfa*** logic.

1. If rising edge of ***rdb*** is detected, ***set_ibfa*** is checked for its logic value. If ***set_ibfa*** is at logic "1", ***reset_ibfa*** is assigned to logic "1". If ***set_ibfa*** is at logic "0", ***reset_ibfa*** is assigned to logic "0".
2. If ***int_reset*** is at logic "1", ***set_ibfa*** is assigned to logic "0".
3. If PLB is operating in mode 1, ***portA*** functions as a strobed input port, ***reset_ibfa*** is at logic "0" and ***stbab*** is at logic "0", ***set_ibfa*** is assigned to logic "1".
4. If PLB is operating in mode 2, ***reset_ibfa*** is at logic "0" and ***stbab*** is at logic "0", ***set_ibfa*** is assigned to logic "1".
5. If PLB is neither operating in mode 1 nor mode 2, ***reset_ibfa*** is checked for its logic value. If ***reset_ibfa*** is at logic "1", ***set_ibfa*** is assigned to logic "0".

The logic to generate ***set_ibfb*** and ***reset_ibfb*** is similar to the logic that generates ***set_ibfa*** and ***reset_ibfa***. However, the logic for generation of ***set_ibfb*** does not involve mode 2 of operation. Figure 6.30 shows the flow diagram for generation of ***set_ibfb*** and ***reset_ibfb***. In Figure 6.30:

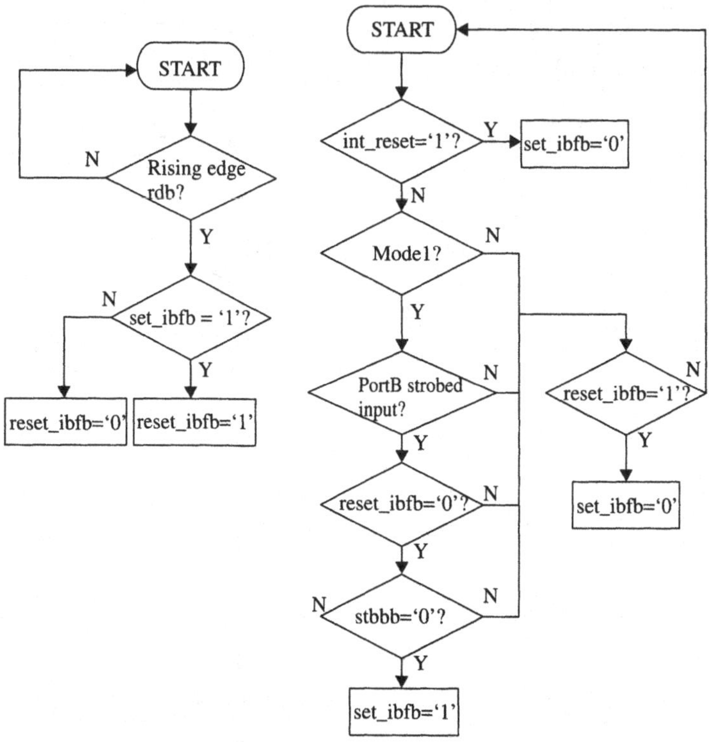

FIGURE 6.30. Diagram showing generation of ***set_ibfb*** and ***reset_ibfb*** logic.

1. If the rising edge of ***rdb*** is detected, ***set_ibfb*** is checked for its logic value. If ***set_ibfb*** is at logic "1", ***reset_ibfb*** is assigned to logic "1". If ***set_ibfb*** is at logic "0", ***reset_ibfb*** is assigned to logic "0".
2. If ***int_reset*** is at logic "1", ***set_ibfb*** is assigned to logic "0".
3. If PLB is operating in mode 1, ***portB*** functions as strobed input, ***reset_ibfb*** is at logic "0" and ***stbbb*** is at logic "0", ***set_ibfb*** is assigned to logic "1".
4. Otherwise, ***reset_ibfb*** is checked for its logic value. If ***reset_ibfb*** is at logic "1", ***set_ibfb*** is assigned to logic "0".

With the generation of the logic ***set_ibfa*** and ***set_ibfb***, these two signals generate ***ibfa*** (***portC***[5]) and ***ibfb*** (***portC***[1]).

Figures 6.11 to 6.30 show the flow diagrams for the generation of logic to create the necessary signals for the operation of PLB. With these logics in place, the final piece of logic required is to multiplex these signals to the ***portA***, ***portB***, and ***portC*** buses. Figure 6.31 shows the multiplexing of signals to ***portA***, Figure 6.32 shows the multiplexing of signals to ***portB*** and Figure 6.33 shows the multiplexing of signals to ***portC***.

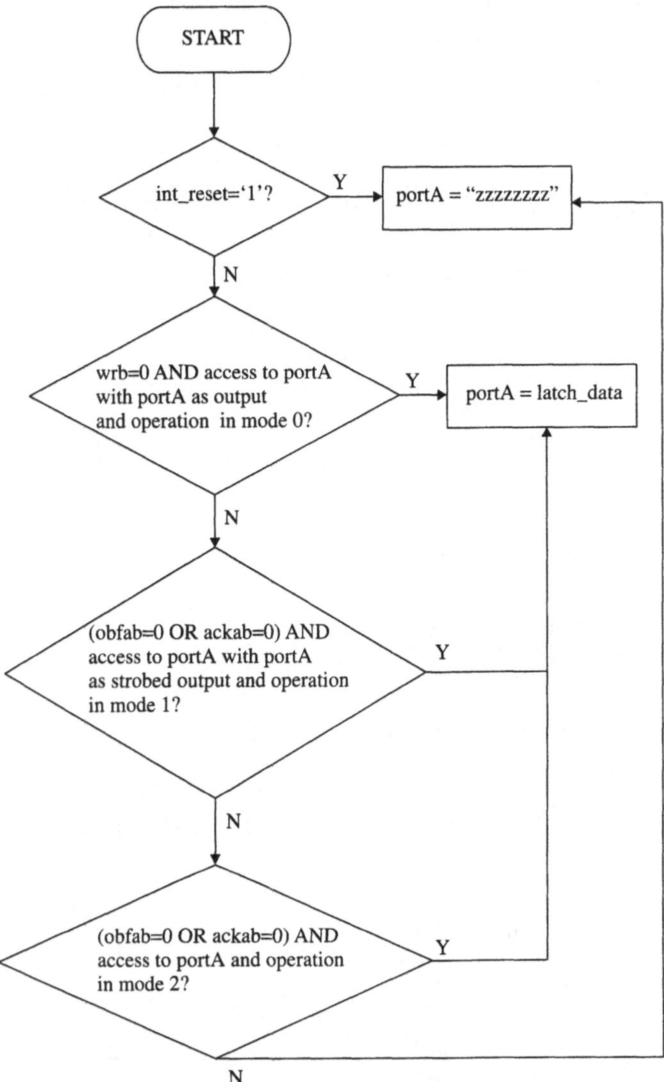

FIGURE 6.31. Diagram showing multiplexing of signals to ***portA***.

6.5 SYNTHESIZABLE VERILOG CODE FOR PROGRAMMABLE PERIPHERAL INTERFACE

Based on the specification and flow diagrams shown in Sections 6.1, 6.2, 6.3, and 6.4, a Verilog code is written for the programmable peripheral interface (as shown in Example 6.1).

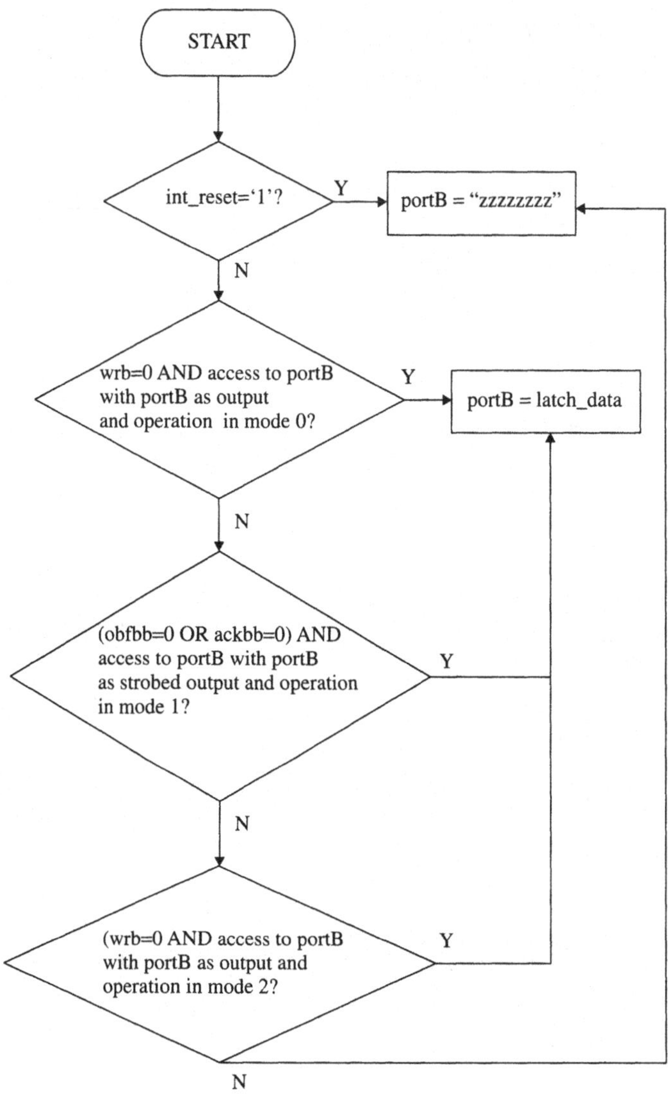

FIGURE 6.32. Diagram showing multiplexing of signals to ***portB***.

Note: There are many ways to build the PLB based on different microarchitectural implementation. The Verilog code of Example 6.1 is based on the microarchitecture explained in Section 6.4. The objective of Example 6.1 is to show the reader how a real-life practical design can be written in synthesizable Verilog.

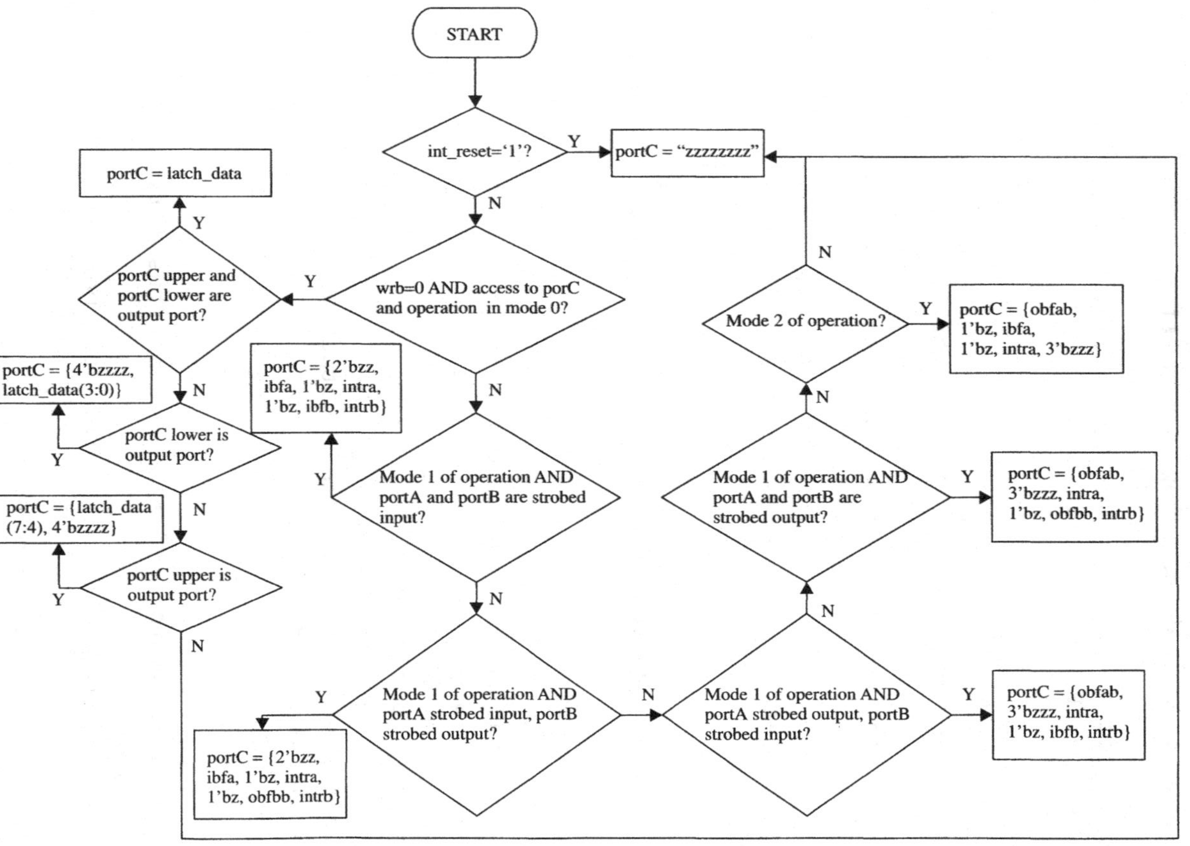

FIGURE 6.33. Diagram showing multiplexing of signals to ***portC***.

Example 6.1 Synthesizable Verilog Code for PLB

```
module ppi (
portA, portB, portC, rdb, wrb, a2, a1, a0,
reset, data);

input rdb, wrb, a2, a1, a0, reset;
inout [7:0] data, portA, portB, portC;

// declaration for control word register
reg [7:0] CWR;

// declaration for status register
reg [7:0] STATUS;

// declaration of internal 3 bit address bus
wire [2:0] address;

assign address = {a2, a1, a0};

// declaration for portA, portB, portC tri-state
// enable signal
reg portAenable, portBenable;
reg [7:0] portCenable;

// declaration for tri-state output bus for data
reg [7:0] out_data;

assign data = out_data;

// declaration for tri-state output bus for portA,
// portB, portC
reg [7:0] out_portA, out_portB, out_portC;

assign portA = out_portA;
assign portB = out_portB;
assign portC = out_portC;

// declaration of integer i;
integer i;

// declaration of internal latching of data
reg [7:0] latch_data;
```

```
// declaration of int_reset
// for internal reset
wire int_reset;

assign int_reset = reset | (~CWR[7]) | (~wrb &
((address == 3'b011) | (address == 3'b111)));

// declaration for latching in of portA and portB and
// portC
reg [7:0] latch_portA_mode0, latch_portB_mode0,
latch_portC;
reg [7:0] latch_portA_mode1_SI, latch_portB_mode1_SI;

// declaration of internal signals
wire stbab, stbbb;
wire ackab, ackbb;
wire out_obfab, out_obfbb;
reg set_obfab, set_obfbb;
wire out_intra, out_intrb;
wire out_ibfa, out_ibfb;
reg set_si_intra, set_so_intra;
reg set_si_intrb, set_so_intrb;

reg set_ibfa, reset_ibfa, set_ibfb, reset_ibfb;

// declaration of wrb_portA, wrb_portB;
wire wrb_portA, wrb_portB;

// declaration of rdb_portA, rdb_portB;
wire rdb_portA, rdb_portB;

// for reset and for writing of data into CWR

always @ (posedge reset or posedge wrb)
begin
    if (reset)
        begin
          // during reset, CWR is set with active flag
          // as active and in mode 0 with all portA, portB
          // and portC as input
            CWR <= 8'b10011110; // bit 0 is not used
            // but default to 0
        end
    else // rising edge wrb
```

Declaration of internal reset to reset PLB.

During reset, ***CWR*** register default to "9e" hex, with all ports as inputs and in mode 0 operation.

```
    begin
        if (address == 3'b011)
    begin
        // write value to CWR
        CWR [7:0] <= data [7:0];
        end
    end
end
```

If positive edge of **wrb** is detected, and the address is pointing to "011" (***CWR*** register), the data from data bus is written into the ***CWR*** register.

```
// for reset and for writing of data into STATUS

always @ (posedge reset or posedge wrb)
begin
    if (reset)
        begin
            // during reset, STATUS is reset to
            // all zeros
            STATUS <= 0;
         end
     else // rising edge wrb
       begin
         if (address == 3'b111)
           begin
             // write value to STATUS
             STATUS [7:0] <= data [7:0];
             end
       end
end
```

During reset, ***STATUS*** register default to "00" hex.

If positive edge of **wrb** is detected, and the address is pointing to "111" (***STATUS*** register), the data from data bus is written into the ***STATUS*** register.

```
// for latching in of data when wrb is at falling
// edge for mode 0 port
// output and also for mode 1
port as strobed output

always @ (posedge int_reset
or negedge wrb)
begin
    if (int_reset)
      latch_data [7:0] <= 8'h00;
    else // falling edge wrb
        latch_data [7:0] <= data [7:0];
end
```

During internal reset, internal registers ***latch_data*** default to "00" hex.

If positive edge of **wrb** is detected, data bus is latched onto the internal registers ***latch_data***.

```
// for latching in of data when rdb is at falling
// edge for mode 0 port
// input and also for mode 1 port as strobed input
```

```
// latch in portA when falling of rdb when in mode 0
// input
// latch in portA when falling edge stbab when in mode
// 1 strobed input
// latch in portB when falling of rdb when in mode 0
// input
// latch in portB when falling edge stbbb when in mode
// 1 strobed input

always @ (negedge rdb or
 posedge int_reset)
begin
    if (int_reset)
      begin
        latch_portA_mode0 [7:0] <= 8'h00;
        latch_portB_mode0 [7:0] <= 8'h00;
      end
    else // falling edge rdb
      begin
        latch_portA_mode0 [7:0]
        <= portA [7:0];
        latch_portB_mode0 [7:0]
        <= portB [7:0];
       end
end

always @ (negedge stbab or posedge int_reset)
begin
    if (int_reset)
        latch_portA_mode1_SI [7:0] <= 8'h00;
    else // falling edge stbab
        latch_portA_mode1_SI [7:0]
        <= portA [7:0];
end

always @ (negedge stbbb or
posedge int_reset)
begin
    if (int_reset)
        latch_portB_mode1_SI [7:0]
        <= 8'h00;
    else // falling edge stbbb
```

During internal reset, internal registers ***latch_portA_mode0*** and ***latch_portB_mode0*** default to "00" hex.

If negative edge of ***rdb*** is detected, the data from ***portA*** bus is latched into ***latch_portA_mode0*** register and data from ***portB*** bus is latched into ***latch_portB_mode0***.

During internal reset, internal registers ***latch_portA_mode1_SI*** default to "00" hex. Otherwise, if falling edge of ***stbab*** is detected, the data from ***portA*** bus is latched into ***latch_portA_mode1_SI***.

```
        latch_portB_mode1_SI
        [7:0] <= portB [7:0];
end
```

During internal reset, internal registers ***latch_portB_mode1_SI*** default to "00" hex. Otherwise if falling edge of ***stbbb*** is detected, the data from ***portB*** bus is latched into ***latch_portB_mode1_SI***.

```
// portC can only be latched
// in at mode 0 input
// latch in of portC during
// falling edge of rdb

always @ (posedge int_reset or negedge rdb)
begin
    if (int_reset)
        latch_portC [7:0]
        <= 8'h00;
    else // falling edge rdb
        latch_portC [7:0]
        <= portC [7:0];
end
```

During internal reset, internal registers ***latch_portC*** default to "00" hex. Otherwise if falling edge of ***rdb*** is detected, the data from ***portC*** bus is latched into ***latch_portC***.

```
// for driving of out_data which is a tristate bus
// for data inout
// out_data is driven when in read mode for mode 0
// or when reading for strobed input for mode 1
// out_data is driven at ~rdb in mode 0 read
// and mode 1 strobed input

always @ (int_reset or rdb or CWR or portAenable or
address or
latch_portA_mode0 or latch_portB_mode0 or portBenable
or portCenable or
latch_portA_mode1_SI or latch_portB_mode1_SI or
latch_portC)
begin
    if (int_reset)
        out_data [7:0] = 8'hzz;
    else if (~rdb & (address == 3'b011))
        out_data [7:0] = CWR[7:0];
    else if (~rdb & (address == 3'b111))
        out_data [7:0] = STATUS [7:0];
    else if (~rdb & (CWR[6:5] == 2'b00) & ~portAenable
            & (address == 3'b000))
        out_data = latch_portA_mode0; // portA mode 0 input
```

See Note A after this Example.

```
    else if (~rdb & (CWR[6:5] == 2'b00) & ~portBenable
            & (address == 3'b001))
        out_data = latch_portB_mode0; // portB mode 0 input
    else if (~rdb & (CWR[6:5] == 2'b00) & (portCenable
            == 8'h00) & (address == 3'b010))
        out_data = latch_portC; // portC mode 0 input
    else if (~rdb & (CWR[6:5] == 2'b00) & (portCenable
            == 8'h0f) & (address == 3'b010))
        out_data = {latch_portC[7:4], 4'hz}; // portC mode 0
                                             // Cupper input
    else if (~rdb & (CWR[6:5] == 2'b00) & (portCenable
            == 8'hf0) & (address == 3'b010))
        out_data = {4'hz, latch_portC[3:0]}; // portC mode 0
                                             // Clower input
    else if (~rdb & (CWR[6:5] == 2'b01) & ~portAenable
    & (address
            == 3'b000))
        out_data = latch_portA_mode1_SI; // portA mode 1
                                         // strobed input
    else if (~rdb & (CWR[6:5] == 2'b01) & ~portBenable
            & (address == 3'b001))
        out_data = latch_portB_mode1_SI; // portB mode
        // 1 strobed input
    else if (~rdb & (CWR[6:5] == 2'b10) & (address
            == 3'b000)) // mode 2 portA is strobed IO
        out_data = latch_portA_mode1_SI; // port A mode 2
                                         // strobed input
    else if (~rdb & (CWR[6:5] == 2'b10) & ~portBenable
            & (address == 3'b001))
        out_data = latch_portB_mode0; // in mode 2 strobed
                                      // input for portB,
                                      // it behaves as
                                      // mode 0 input
    else
        out_data = 8'hzz;
end

// for generation of portA, portB, portC enable
// signal to control tri-state buffer

always @ (CWR or int_reset)
begin
    if (int_reset)
        begin
        // after reset, all portA, portB, portC are input
```

During internal reset, ***portAenable, portBenable*** and ***portCenable*** defaults to zero, which means all ports default as input.

```
            portAenable = 0;
            portBenable = 0;
            portCenable = 8'h00;
         end
      else
         begin
            if (CWR[6:5] == 2'b00)───► See Note B after
                                       this Example.
               begin
                  // this is mode 0
                  if (~CWR[4])
                     portCenable [7:4] = 4'hf; // port C upper
                                               // is output
                  else
                     portCenable [7:4] = 4'h0; // port C upper
                                               // is input
                  if (~CWR[3])
                     portCenable [3:0] = 4'hf; // port C lower
                                               // is output
                  else
                     portCenable [3:0] = 4'h0; // port C lower
                                               // is input
                  if (~CWR[2])
                     portBenable = 1; // port B is output
                  else
                     portBenable = 0; // port B is input

                  if (~CWR[1])
                     portAenable = 1; // port C is output
                  else
                     portAenable = 0; // port C is input
               end
            else if (CWR[6:5] == 2'b01)
            begin
               // this is mode 1
               if (~CWR[2])
                  portBenable = 1; // port B is strobed
                                   // output
               else
                  portBenable = 0; // portB is strobed
                                   // input
               if (~CWR[1])
                  portAenable = 1; // port A is strobed
                                   // output
               else
                  portAenable = 0;  // port A is strobed input
```

```
if (CWR[2:1] == 2'b00) // portA and B is
                        // strobed output
   portCenable = 8'b10001011; // this
   // translates
                 // to portC7 is output,
                 // portC6 is input,
                 // portC4 and portC5 is
                 // not used but default
                 // to input, portC3 is
                 // output, portC2 is
                 // input, portC1 is
                 // output and portC0 is
                 // output
else if (CWR[2:1] == 2'b01) // portA strobed
                 // input, portB strobed
                 // output
   portCenable = 8'b00101011; // this translates
                 // to portC7 and portC6 not
                 // used but default to
                 // input, portC5 is
                 // output, portC4 is
                 // input, portC3 is
                 // output, portC2 is
                 // input, portC1 is
                 // output, portC0 is
                 // output
else if (CWR[2:1] == 2'b10) // portA strobed
                 // output, portB strobed
                 // input
   portCenable = 8'b10001011; // this translates
                 // to portC7 is output,
                 // portC6 is input,
                 // portC5 and portC4
                 // is not used but
                 // default to input,
                 // portC3 is output,
                 // portC2 is input,
                 // portC1 is output
                 // and portC0 is output
else // (CWR[2:1] == 2'b11)  - portA strobed
                 // input, portB strobed
                 // input
   portCenable = 8'b00101011; // this translates
                 // to portC7 and portC6
                 // not used but default
```

```
                         // to input, portC5 is
                         // output, portC4 is
                         // input, portC3 is
                         // output, portC2 is
                         // input, portC1 is
                         // output and portC0
                         // is output
               end
            else if (CWR[6:5] == 2'b10)
               begin
                  // this is mode 2
                  if (~CWR[2])
                     portBenable = 1; // portB is
                     // output in mode 0
                  else
                     portBenable = 0;

                  // in this mode 2, port A is
                  // bidirectional and
                  // portC is used as handshake signal
                  portCenable = 8'b10101000; // this
                  // translates to
                         // portC7 is output, portC6
                         // is input, portC5 is
                         // output, portC4 is input
                         // portC3 is output, portC2,
                         // portC1 and portC0 is not
                         // used but default to
                         // input
            end
        end
end


// generation of output obfab &
// obfbb


always @ (posedge wrb or negedge ackab)
begin
   if (~ackab) // falling edge ackab
      set_obfab <= 0;
   else if (address == 3'b000)
   // posedge wrb
      set_obfab <= 1;
end
```

If negative edge of ***ackab*** is detected, ***set_obfab*** is driven to logic "0". If rising edge of ***wrb*** is detected when address is "000", ***set_obfab*** is driven to logic "1".

```
always @ (posedge wrb or negedge ackbb)
begin
   if (~ackbb) // falling edge ackbb
      set_obfbb <= 0;
   else if (address == 3'b001)
   // posedge wrb
      set_obfbb <= 1;
end
```

If negative edge of ***ackbb*** is detected, ***set_obfbb*** is driven to logic "0". If rising edge of ***wrb*** is detected when address is "001", ***set_obfbb*** is driven to logic "1".

```
assign out_obfab = ~(set_obfab &
(((CWR[6:5] == 2'b01) & portAenable)
    | (CWR[6:5] == 2'b10)) &
    (address != 3'b011)
         & (address != 3'b111));

assign out_obfbb = ~(set_obfbb & ((CWR[6:5] == 2'b01)
& portBenable)
         & (address != 3'b011) & (address != 3'b111)) ;

// generation of input ackab
// needed when mode 1 and portA is strobed output or
// when mode 2

assign ackab = (((CWR[6:5] == 2'b01) & portAenable &
portC[6]) | (CWR[6:5] == 2'b10) & portC[6]);

// generation of input ackbb
// needed when in mode 1 and portB is strobed output

assign ackbb = ((CWR[6:5] == 2'b01) & portBenable &
portC[2]);

// generation of output intra
// needed when in mode 1 or when mode 2

// for portA in mode 1 strobed output, falling edge of
// wrb reset intra

assign wrb_portA = (address == 3'b000) & wrb; // <-
// wrb_portA is used
                  // because for synthesis, synthesis
                  // tools cannot match simulation
                  // if address is used in the
                  // edge declaration of always
                  // statement
```

```
always @ (negedge wrb_portA or posedge ackab)
begin
   if (~wrb_portA) //
   // falling edge wrb at
      portA set_so_intra <= 0;
   else // rising edge ackab
      set_so_intra <= 1;
end
```

set_so_intra is driven to logic "1" when rising edge of **ackab** is detected. If falling edge of **wrb_portA** is detected, **set_so_intra** is driven to logic "0".

```
// for portA in mode 1 strobed input, falling edge rdb
// reset intra

assign rdb_portA = (address == 3'b000) & rdb;

always @ (posedge stbab or negedge rdb_portA)
begin
   if (~rdb_portA) // falling
   // edge rdb at portA
      set_si_intra <= 0;
   else // rising edge stbab
      set_si_intra <= 1;
end
```

set_si_intra is driven to logic '1' when rising edge of **stbab** is detected. If falling edge of **rdb_portA** is detected, **set_si_intra** is driven to logic '0'.

```
assign out_intra = int_reset ?
           0 : (portAenable &
           (CWR[6:5] == 2'b01)
           & STATUS[2] ) ?
           set_so_intra :
           (~portAenable &
           (CWR[6:5] == 2'b01) &
           STATUS[0] ) ?
           set_si_intra : ((CWR[6:5] == 2'b10) &
           STATUS[4]
           & STATUS[5]) ? (set_so_intra |
           set_si_intra) : 0 ;
```

Assigning **out_intra** based on different logical conditions.

```
// generation of output intrb
// needed in mode 1


// for portB in mode 1 strobed output, falling edge of
// wrb reset intrb


assign wrb_portB = (address == 3'b001) & wrb;
```

```
always @ (negedge wrb_portB or posedge ackbb)
begin
   if (~wrb_portB) //
   // falling edge wrb at portB
      set_so_intrb <= 0;
   else // rising edge ackbb
      set_so_intrb <= 1;
end
```

> ***set_so_intrb*** is driven to logic '1' when rising edge of ***ackbb*** is detected. If falling edge of ***wrb_portB*** is detected, ***set_so_intrb*** is driven to logic '0'.

```
// for portB in mode 1 strobed output, falling edge
// rdb reset intrb

assign rdb_portB = (address == 3'b001) & rdb;

always @ (posedge stbbb or negedge rdb_portB)
begin
   if (~rdb_portB) //
   // falling edge rdb at portB
      set_si_intrb <= 0;
   else // rising edge stbbb
      set_si_intrb <= 1;
end
```

> ***set_si_intrb*** is driven to logic "1" when rising edge of ***stbbb*** is detected. If falling edge of ***rdb_portB*** is detected, ***set_si_intrb*** is driven to logic "0".

```
assign out_intrb = int_reset ? 0 : (portBenable &
            (CWR[6:5] == 2'b01) & STATUS[3]) ?
            set_so_intrb : (~portBenable & (CWR[6:5] ==
            2'b01) &
               STATUS[1]) ?
            set_si_intrb : 0;
```

> Assigning ***out_intrb*** based on different logical conditions.

```
// generation of input stbab
// needed when in mode 1 and portA is strobed input
// or when in mode 2

assign stbab = (((CWR[6:5] == 2'b01) & ~portAenable) |
(CWR[6:5] == 2'b10)) ? portC[4] : 1'b1;

// generation of input stbbb
// needed when in mode 1 and portB is strobed input

assign stbbb = ((CWR[6:5] == 2'b01) & ~portBenable)
? portC[2] : 1'b1;

// generation of output ibfa
// needed when in mode 1 and portA is strobed input
// or mode 2
```

```
always @ (posedge rdb)
begin
   if (set_ibfa)
      reset_ibfa <= 1;
   else
      reset_ibfa <= 0;
end
```

reset_ibfa is set when rising edge **rdb** is detected with **set_ibfa** at logic "1". **reset_ibfa** is reset when rising edge **rdb** is detected with **set_ibfa** at logic "0".

```
always @ (CWR or portAenable or reset_ibfa or stbab or
out_ibfa or int_reset)
begin
   if (int_reset)
      set_ibfa = 0;
   else if ((CWR[6:5] == 2'b01) & ~portAenable &
   ~reset_ibfa & ~stbab) // portA input mode 1
      set_ibfa = 1;
   else if ((CWR[6:5] == 2'b10) & ~reset_ibfa
   & ~stbab) // portA mode 2
      set_ibfa = 1;
   else if (reset_ibfa)
      set_ibfa = 0;
end

assign out_ibfa = set_ibfa;

// generation of output ibfb
// needed when in mode 1 and portB is strobed input
always @ (posedge rdb)
begin
   if (set_ibfb)
      reset_ibfb <= 1;
   else
      reset_ibfb <= 0;
end
```

reset_ibfa is set when rising edge **rdb** is detected with **set_ibfa** at logic '1'. **reset_ibfa** is reset when rising edge **rdb** is detected with **set_ibfa** at logic '0'.

```
always @ (CWR or portBenable or reset_ibfb or stbbb or
out_ibfb or int_reset)
begin
   if (int_reset)
      set_ibfb = 0;
   else if ((CWR[6:5] == 2'b01) & ~portBenable &
   ~reset_ibfb & ~stbbb) // portB input mode 1
      set_ibfb = 1;
```

```
   else if (reset_ibfb)
      set_ibfb = 0;
end

assign out_ibfb = set_ibfb;

// writing to portA

always @ (int_reset or wrb or address or CWR or
portAenable or latch_data or out_obfab or ackab)
begin
   if (int_reset)
      out_portA [7:0] = 8'bzzzzzzzz;
   else if (~wrb & (address == 3'b000) & (CWR[6:5] ==
   2'b00) & portAenable) // mode 0
      out_portA [7:0] = latch_data [7:0]; // writing
      // to portA
   else if ((~out_obfab | ~ackab) & (address ==
   3'b000) & (CWR[6:5] == 2'b01) & portAenable) //
   // mode 1
      out_portA [7:0] = latch_data [7:0]; // writing
      // to portA
   else if ((~out_obfab | ~ackab) & (address ==
   3'b000) & (CWR[6:5] == 2'b10)) // mode 2
      out_portA [7:0] = latch_data [7:0]; // writing
      // to portA
   else
      out_portA [7:0] = 8'bzzzzzzzz;
end
```

Data is written to ***portA*** when ***portA*** functions as output port in mode 0, strobed output port in mode 1 and strobed I/O in mode 2.

```
// writing to portB

always @ (int_reset or wrb or
address or CWR or portBenable or
latch_data or out_obfbb or ackbb)
begin
   if (int_reset)
      out_portB [7:0] = 8'bzzzzzzzz;
   else if (~wrb & (address == 3'b001) & (CWR[6:5] ==
   2'b00) & portBenable) // mode 0
      out_portB [7:0] = latch_data [7:0]; // writing
      // to portB
   else if ((~out_obfbb | ~ackbb) & (address ==
   3'b001) & (CWR[6:5] == 2'b01) & portBenable) //
   // mode 1
```

Data is written to ***portB*** when ***portB*** functions as output port in mode 0, strobed output port in mode 1 and output port in mode 2.

```
      out_portB [7:0] = latch_data [7:0]; // writing
      // to portB
   else if (~wrb & (address == 3'b001) & (CWR[6:5]
   == 2'b10) & portBenable) // mode 2 port B output
      out_portB [7:0] = latch_data [7:0];
   else
      out_portB [7:0] = 8'bzzzzzzzz;
end


// writing to portC


always @ (int_reset or wrb or
address or CWR or portCenable or
latch_data or out_ibfa or out_
intra or out_ibfb or out_intrb
or out_obfab or out_obfbb or
portAenable or portBenable)
begin
   if (int_reset)
      out_portC [7:0] = 8'bzzzzzzzz;
   else if (~wrb & (address == 3'b010) &
   (CWR[6:5] == 2'b00)) //mode 0
      begin
         if (portCenable == 8'hff)
            out_portC [7:0] = latch_data [7:0];
         else if (portCenable == 8'h0f)
            out_portC [7:0] = {4'bzzzz, latch_data
            [3:0]};
         else if (portCenable == 8'hf0)
            out_portC [7:0] = {latch_data [7:4],
            4'bzzzz};
         else
            out_portC [7:0] = 8'bzzzzzzzz;
      end
   else if (CWR[6:5] == 2'b01) // this is mode 1
      begin
         // in mode 1, for portA strobed input,
         // portB strobed input
         // portC[7] not used
         // portC[6] not used
         // portC[5] output ibfa
         // portC[4] input stbab
         // portC[3] output intra
         // portC[2] input stbbb
```

Data is written to ***portC*** when ***portC*** functions as output port in mode 0. In mode 0, ***portC*** is separated into ***portC***[7:4] and ***portC***[3:0].

Control signals are written to ***portC*** in mode 1 operation.

```
// portC[1] output ibfb
// portC[0] output intrb
if (~portAenable & ~portBenable) // this is
// portA,
                    // portB as strobed input
   out_portC[7:0] = {2'bzz, out_ibfa, 1'bz,
      out_intra, 1'bz, out_ibfb,
      out_intrb};
// in mode 1, for portA strobed input, portB
// strobed output
// portC[7] not used
// portC[6] not used
// portC[5] output ibfa
// portC[4] input stbab
// portC[3] output intra
// portC[2] input ackbb
// portC[1] output obfbb
// portC[0] output intrb
else if (~portAenable & portBenable)
   out_portC[7:0] = {2'bzz, out_ibfa, 1'bz,
         out_intra, 1'bz, out_obfbb,
         out_intrb};
// in mode 1, for portA strobe output and portB
// strobed input
// portC[7] output obfab
// portC[6] input ackab
// portC[5] not used
// portC[4] not used
// portC[3] output intra
// portC[2] input stbbb
// portC[1] output ibfb
// portC[0] output intrb
else if (portAenable & ~portBenable)
   out_portC[7:0] = {out_obfab, 3'bzzz,
      out_intra, 1'bz, out_ibfb,
      out_intrb};
// in mode 1, for portA strobe output and portB
// strobed output
// portC[7] output obfab
// portC[6] input ackab
// portC[5] not used
// portC[4] not used
// portC[3] out_intra
// portC[2] input ackbb
// portC[1] output obfbb
```

```
            // portC[0] output intrb
            else // portAenable and portBenable is high
               out_portC[7:0] = {out_obfab, 3'bzzz,
                  out_intra, 1'bz, out_obfbb,
                  out_intrb};
         end
else if (CWR[6:5] == 2'b10) // this is mode 2
         // portC[7] output obfab
         // portC[6] input ackab
         // portC[5] output ibfa
         // portC[4] input stbab
         // portC[3] output intra
         // portC[2] not used
         // portC[1] not used
         // portC[0] not used
            out_portC[7:0] = {out_obfab, 1'bz, out_ibfa, 1'bz,
               out_intra, 3'bzzz};
      else
               out_portC[7:0] = 8'bzzzzzzzz;
end

endmodule
```

Control signals are written to ***portC*** in mode 2 operation.

Note A:

a. During internal reset, ***out_data*** is tri-stated.
b. If ***rdb*** is at logic low and address is pointing to "011" (***CWR*** register), the contents of ***CWR*** register are read onto ***out_data***.
c. If ***rdb*** is at logic low and address is pointing to "111" (***STATUS*** register), the contents of ***STATUS*** register are read onto ***out_data***.
d. If ***rdb*** is at logic low, bits 6 and 5 of ***CWR*** register have a value of "00" (operation in mode 0), ***portAenable*** is at logic low (***portA*** functions as input port), and address is pointing to "000" (***portA***), the contents of internal register ***latch_portA_mode0*** are read onto ***out_data***.
e. If ***rdb*** is at logic low, bits 6 and 5 of ***CWR*** register have a value of "00" (operation in mode 0), ***portBenable*** is at logic low (***portB*** functions as input port) and address is pointing to "001" (***portB***), the contents of internal register ***latch_portB_mode0*** are read onto ***out_data***.
f. If ***rdb*** is at logic low, bits 6 and 5 of ***CWR*** register have a value of "00" (operation in mode 0), ***portCenable*** has a value of "00000000" (***portC*** functions as input port) and address is pointing to "010" (***portC***), the contents of internal register ***latch_portC*** are read onto ***out_data***.
g. If ***rdb*** is at logic low, bits 6 and 5 of ***CWR*** register have a value of "00" (operation in mode 0), ***portCenable*** has a value of "00001111" (***portC***

upper functions as input port and ***portC*** lower functions as output port) and address is pointing to "010" (***portC***), the contents of internal register ***latch_portC*** bits 7 to 4 are read onto ***out_data*** bits 7 to 4. The remaining bits 3 to 0 of ***out_data*** are tri-stated.

h. If ***rdb*** is at logic low, bits 6 and 5 of ***CWR*** register have a value of "00" (operation in mode 0), ***portCenable*** has a value of "11110000" (***portC*** lower functions as input port and ***portC*** upper functions as output port) and address is pointing to "010" (***portC***), the contents of internal register ***latch_portC*** bits 3 to 0 are read onto ***out_data*** bits 3 to 0. The remaining bits 7 to 4 of ***out_data*** are tri-stated.

i. If ***rdb*** is at logic low, bits 6 and 5 of ***CWR*** register have a value of "01" (operation in mode 1), ***portAenable*** is at logic low (***portA*** functions as strobed input port) and address is pointing to "000" (***portA***), the contents of internal register ***latch_portA_mode1_SI*** are read onto ***out_data***.

j. If ***rdb*** is at logic low, bits 6 and 5 of ***CWR*** register have a value of "01" (operation in mode 1), ***portBenable*** is at logic low (***portB*** functions as strobed input port) and address is pointing to "001" (***portB***), the contents of internal register ***latch_portB_mode1_SI*** are read onto ***out_data***.

k. If ***rdb*** is at logic low, bits 6 and 5 of ***CWR*** register have a value of "10" (operation in mode 2) and address is pointing to "000" (***portA***), the contents of internal register ***latch_portA_mode1_SI*** are read onto ***out_data***.

l. If ***rdb*** is at logic low, bits 6 and 5 of ***CWR*** register have a value of "10" (operation in mode 2), ***portBenable*** is at logic low (***portB*** functions as input port) and address is pointing to "001" (***portB***), the contents of internal register ***latch_portB_mode0*** are read onto ***out_data***.

m. If neither of the conditions mentioned are met, ***out_data*** is tri-stated.

Note B:

a. Bits 6 and 5 of ***CWR*** register are checked for their value. Assume ***CWR*** bits 6 and 5 are both at logic "0" (mode 0 operation)
   i. If bit 4 of ***CWR*** register is at logic "0", bits 7 to 4 of ***portCenable*** are driven to logic "1". If bit 4 of ***CWR*** register is at logic "1", bits 7 to 4 of ***portCenable*** are driven to logic "0".
   ii. If bit 3 of ***CWR*** register is at logic "0", bits 3 to 0 of ***portCenable*** are driven to logic "1". If bit 3 of ***CWR*** register is at logic "1", bits 3 to 0 of ***portCenable*** are driven to logic "0".
   iii. If bit 2 of ***CWR*** register is at logic "0", ***portBenable*** is driven to logic "1". If bit 2 of ***CWR*** register is at logic "1", ***portBenable*** is driven to logic "0".

iv. If bit 1 of ***CWR*** register is at logic "0", ***portAenable*** is driven to logic "1". If bit 1 of ***CWR*** register is at logic "1", ***portAenable*** is driven to logic "0".

b. Bits 6 and 5 of ***CWR*** register are checked for their value. Assume ***CWR[6:5]*** is at logic "01" (mode 1 operation)
   i. If bit 2 of ***CWR*** register is at logic "0", ***portBenable*** is driven to logic "1". If bit 2 of ***CWR*** register is at logic "1", ***portBenable*** is driven to logic "0".
   ii. If bit 1 of ***CWR*** register is at logic "0", ***portAenable*** is driven to logic "1". If bit 1 of ***CWR*** register is at logic "1", ***portAenable*** is driven to logic "0".
   iii. If ***CWR[2:1]*** register is at logic "00" (***portA*** and ***portB*** function as strobed output), ***portCenable*** is driven to logic "10001011". This translates to ***portC***[7] is output, ***portC***[6:4] is input, ***portC***[3] is output, ***portC***[2] is input, and ***portC***[1:0] is output.
   iv. If ***CWR[2:1]*** register is at logic "01" (***portA*** functions as strobed input and ***portB*** functions as strobed output), ***portCenable*** is driven to logic "00101011". This translates to ***portC***[7:6] is input, ***portC***[5] is output, ***portC***[4] is input, ***portC***[3] is output, ***portC***[2] is input, and ***portC***[1:0] is output.
   v. If ***CWR[2:1]*** register is at logic "10" (***portA*** functions as strobed output and ***portB*** functions as strobed input), ***portCenable*** is driven to logic "10001011". This translates to ***portC***[7] is output, ***portC***[6:4] is input, ***portC***[3] is output, ***portC***[2] is input, and ***portC***[1:0] is output.
   vi. If ***CWR[2:1]*** register is at logic "11" (***portA*** and ***portB*** function as strobed input), ***portCenable*** is driven to logic "00101011". This translates to ***portC***[7:6] is input, ***portC***[5] is output, ***portC***[4] is input, ***portC***[3] is output, ***portC***[2] is input, and ***portC***[1:0] is output.

c. Bits 6 and 5 of ***CWR*** register are checked for their value. Assume ***CWR[6:5]*** is at logic "10" (mode 2 operation)
   i. If bit 2 of ***CWR*** register is at logic "0", ***portBenable*** is driven to logic "1". If bit 2 of ***CWR*** register is at logic "1", ***portBenable*** is driven to logic "0".
   ii. ***PortC*** functions as strobed I/O. ***PortCenable*** is driven to logic "10101000". This translates to ***portC***[7] is output, ***portC***[6] is input, ***portC***[5] is output, ***portC***[4] is input, ***portC***[3] is output, and ***portC***[2:0] is input.

6.6 SIMULATION FOR PROGRAMMABLE PERIPHERAL INTERFACE DESIGN

The Verilog code shown in Example 6.1 implements the functionality explained in Section 6.4. To verify the Verilog code has the correct functionality, the code is simulated. A total of 13 test benches are written to simulate the different functionalities of the PLB design:

1. mode 0 operation with ***portA***, ***portB***, and ***portC*** as input port
2. mode 0 operation with ***portA***, ***portB***, and ***portC*** as output port
3. mode 0 operation with ***portA***, ***portB***, ***portC*** lower as input port, and ***portC*** upper as output port
4. mode 0 operation with ***portA***, ***portB***, ***portC*** upper as input port, and ***portC*** lower as output port
5. writing data to and reading data from ***STATUS*** and ***CWR*** register
6. mode 1 operation with ***portA*** and ***portB*** as strobed input
7. mode 1 operation with ***portA*** as strobed input and ***portB*** as strobed output
8. mode 1 operation with ***portA*** as strobed output and ***portB*** as strobed input
9. mode 1 operation with ***portA*** and ***portB*** as strobed output
10. mode 2 operation with ***portA*** as strobed I/O and ***portB*** as input
11. mode 2 operation with ***portA*** as strobed I/O and ***portB*** as output
12. mode 1 operation with ***portA*** and ***portB*** as strobed input (***STATUS*** registers at logic "0")
13. mode 2 operation with ***portA*** as strobed I/O and ***portB*** as output (***STATUS*** registers at logic "0")

Sections 6.6.1 to 6.6.12 show the test bench that can be used to simulate the PLB design for different functionalities. Waveforms of the simulation results are shown to explain the functionality of the design.

6.6.1 Simulation for Mode 0 Operation with *PortA, PortB,* and *PortC* as Input and Output

In mode 0 operation, ***portA*** and ***portB*** can function as either input ports or output ports. ***PortC*** is divided into ***portC*** upper and ***portC*** lower, each able to function as input port or output port. For this simulation, all ports are set up to function as input ports. The ports are then reconfigured to function as output ports. Example 6.2 shows the Verilog code that can be used to simulate the PLB design.

Example 6.2 Simulation with *PortA*, *PortB*, and *PortC* as Input and Output in Mode 0 Operation

```
module ppi_tb ();

wire [7:0] tb_portA, tb_portB, tb_portC;
reg tb_rdb, tb_wrb, tb_reset;
```

```
wire [7:0] tb_data;
wire tb_a2, tb_a1, tb_a0;

reg [7:0] drive_portA, drive_portB, drive_portC,
drive_data, temp_data;

parameter cycle = 100;

assign tb_portA = drive_portA;
assign tb_portB = drive_portB;
assign tb_portC = drive_portC;
assign tb_data = drive_data;

reg [2:0] address;
assign tb_a2 = address [2];
assign tb_a1 = address [1];
assign tb_a0 = address [0];

initial
begin
     // for reset
     drive_portA = 8'hzz;
     drive_portB = 8'hzz;
     drive_portC = 8'hzz;
     tb_rdb = 1;
     tb_wrb = 1;
     address = 0;
     drive_data = 8'hzz;
     tb_reset = 0;

     #cycle;

     task_reset;

     // testing for mode 0 with all portA, portB,
     // portCU, portCL
     // input

     temp_data = 8'b10011110;
     CWR_write(temp_data);
     drive_data = 8'hzz;

     // read from portA
     address = 0;
     drive_portA = 8'ha5;
     read_port;
     drive_portA = 8'hzz;
```

```
    // read portB
    address = 1;
    drive_portB = 8'h35;
    read_port;
    drive_portB = 8'hzz;

    // read portC
    address = 2;
    drive_portC = 8'h98;
    read_port;
    drive_portC = 8'hzz;

    // change portA to output
    temp_data = 8'b10011100;
    CWR_write(temp_data);
    drive_data = 8'hzz;

    // write to portA
    address = 0;
    drive_data = 8'hbc;
    write_port;
    drive_data = 8'hzz;

    // change portB to output
    temp_data = 8'b10011000;
    CWR_write(temp_data);
    drive_data = 8'hzz;

    // write to portB
    address = 1;
    drive_data = 8'h67;
    write_port;
    drive_data = 8'hzz;

    // change portC to output
    temp_data = 8'b10000000;
    CWR_write(temp_data);
    drive_data = 8'hzz;

    // write to portC
    address = 2;
    drive_data = 8'h32;
    write_port;
    drive_data = 8'hzz;

end
```

```
task write_port;
begin
     tb_wrb = 1;
     tb_rdb = 1;
     #cycle;
     tb_wrb = 0;
     #cycle;
     tb_wrb = 1;
     #cycle;
end
endtask

task read_port;
begin
     tb_wrb = 1;
     tb_rdb = 1;
     #cycle;
     tb_rdb = 0;
     #cycle;
     tb_rdb = 1;
     #cycle;
end
endtask

task CWR_write;
input [7:0] temp_data;
begin
     tb_reset = 0;
     address = 3'b011;
     tb_rdb = 1;
     tb_wrb = 1;
     #cycle;
     drive_data = temp_data;
     tb_wrb = 0;
     #cycle;
     tb_wrb = 1;
     #cycle;
end
endtask

task task_reset;
begin
     tb_reset = 0;
     #cycle;
     tb_reset = 1;
```

```
      #cycle;
      tb_reset = 0;
      #cycle;
end
endtask

ppi ppi_inst (.portA(tb_portA), .portB(tb_portB),
.portC(tb_portC), .rdb(tb_rdb),
.wrb(tb_wrb), .a2(tb_a2), .a1(tb_a1), .a0(tb_a0),
.reset(tb_reset),
.data(tb_data));

endmodule
```

Referring to Figure 6.34:

a. In the circle marked 1, ***reset*** occurs to reset the ***CWR*** register to "9e" and ***STATUS*** register to "0".
b. In the circle marked 2, ***wrb*** pulses low, causing a write to address "011" (***CWR*** register). The data on ***data[7:0]*** bus is "9e". This data is written into ***CWR*** register ("9e" in ***CWR*** register configures the PLB to function in mode 0 operation with ***portA***, ***portB*** and ***portC*** as input ports).
c. In the circle marked 3, ***rdb*** pulses low, causing a read from address "000". Because the PLB is operating in mode 0 with ***portA*** as input, the data at ***portA*** ("a5") is read to ***data[7:0]*** bus.

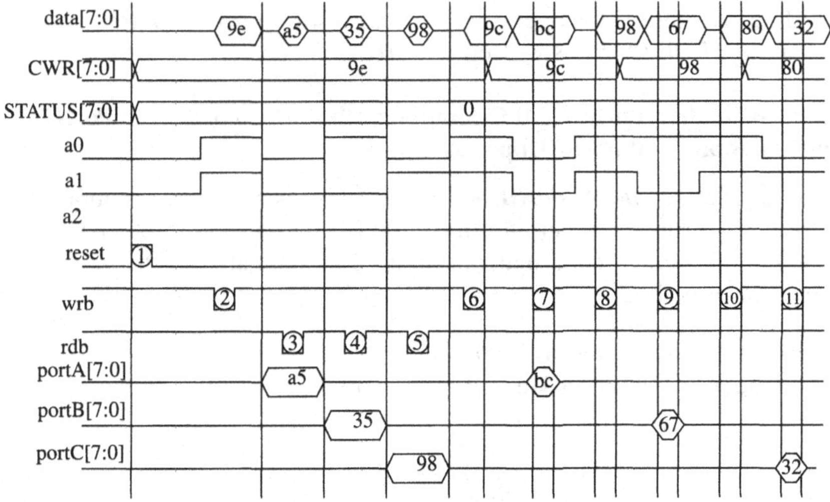

FIGURE 6.34. Diagram showing simulation results of test bench Example 6.2.

d. In the circle marked 4, ***rdb*** pulses low, causing a read from address "001". Because the PLB is operating in mode 0 with ***portB*** as input, the data at ***portB*** ("35") is read to ***data[7:0]*** bus.

e. In the circle marked 5, ***rdb*** pulses low, causing a read from address "010". Because the PLB is operating in mode 0 with ***portC*** as input, the data at ***portC*** ("98") is read to ***data[7:0]*** bus.

f. In the circle marked 6, wrb pulses low, causing a write to address "011" (***CWR*** register). The data on ***data[7:0]*** bus is "9c". This data is written into ***CWR*** register ("9c" in ***CWR*** register configures the PLB to function in mode 0 operation with ***portA*** as output port and ***portB***, ***portC*** as input ports).

g. In the circle marked 7, wrb pulses low, causing a write to address "000". Because the PLB is operating in mode 0 with ***portA*** as output, the data at ***data[7:0]*** ("bc") is written to ***portA***.

h. In the circle marked 8, wrb pulses low, causing a write to address "011" (***CWR*** register). The data on ***data[7:0]*** bus is "98". This data is written into ***CWR*** register ("98" in ***CWR*** register configures the PLB to function in mode 0 operation with ***portA*** and ***portB*** as output port while ***portC*** as input port).

i. In the circle marked 9, wrb pulses low, causing a write to address "001". Because the PLB is operating in mode 0 with ***portB*** as output, the data at ***data[7:0]*** ("67") is written to ***portB***.

j. In the circle marked 10, wrb pulses low, causing a write to address "011" (***CWR*** register). The data on ***data[7:0]*** bus is "80". This data is written into ***CWR*** register ("80" in ***CWR*** register configures the PLB to function in mode 0 operation with ***portA***, ***portB*** and ***portC*** as output port).

k. In the circle marked 11, wrb pulses low, causing a write to address "010". Because the PLB is operating in mode 0 with ***portC*** as output, the data at ***data[7:0]*** ("32") is written to ***portC***.

6.6.2 Simulation for Mode 0 Operation with *PortA*, *PortB*, and *PortC* Lower as Input, and *PortC* Upper as Output

In this simulation, ***portA***, ***portB***, and ***portC*** lower is configured as input ports whereas ***portC*** upper as output port. The PLB operates in mode 0. Example 6.3 shows the Verilog code that can be used to simulate the PLB design for correct functionality.

Example 6.3 Simulation with *PortA*, *PortB*, and *PortC* Lower as Input and *PortC* Upper as Output in Mode 0 Operation

```
module ppi_tb ();


wire [7:0] tb_portA, tb_portB, tb_portC;
reg tb_rdb, tb_wrb, tb_reset;
```

```
wire [7:0] tb_data;
wire tb_a2, tb_a1, tb_a0;

reg [7:0] drive_portA, drive_portB, drive_portC,
drive_data, temp_data;

parameter cycle = 100;

assign tb_portA = drive_portA;
assign tb_portB = drive_portB;
assign tb_portC = drive_portC;
assign tb_data = drive_data;

reg [2:0] address;
assign tb_a2 = address [2];
assign tb_a1 = address [1];
assign tb_a0 = address [0];

initial
begin
     // for reset
     drive_portA = 8'hzz;
     drive_portB = 8'hzz;
     drive_portC = 8'hzz;
     tb_rdb = 1;
     tb_wrb = 1;
     address = 0;
     drive_data = 8'hzz;
     tb_reset = 0;

     #cycle;

     task_reset;

     // testing for mode 0 with all portA, portB
     // input, portCU
     // output, portCL input

     temp_data = 8'b10001110;
     CWR_write(temp_data);
     drive_data = 8'hzz;

     // read from portA
     address = 0;
     drive_portA = 8'ha5;
```

```
     read_port;
     drive_portA = 8'hzz;

     // read portB
     address = 1;
     drive_portB = 8'h35;
     read_port;
     drive_portB = 8'hzz;

     // read portC
     address = 2;
     drive_portC = 8'h98;
     read_port;
     drive_portC = 8'hzz;

     // write to portCupper
     address = 2;
     drive_data = 8'h32;
     write_port;
     drive_data = 8'hzz;

end

task write_port;
begin
     tb_wrb = 1;
     tb_rdb = 1;
     #cycle;
     tb_wrb = 0;
     #cycle;
     tb_wrb = 1;
     #cycle;
end
endtask

task read_port;
begin
     tb_wrb = 1;
     tb_rdb = 1;
     #cycle;
     tb_rdb = 0;
     #cycle;
     tb_rdb = 1;
     #cycle;
end
endtask
```

```
task CWR_write;
input [7:0] temp_data;
begin
     tb_reset = 0;
     address = 3'b011;
     tb_rdb = 1;
     tb_wrb = 1;
     #cycle;
     drive_data = temp_data;
     tb_wrb = 0;
     #cycle;
     tb_wrb = 1;
     #cycle;
end
endtask

task task_reset;
begin
     tb_reset = 0;
     #cycle;
     tb_reset = 1;
     #cycle;
     tb_reset = 0;
     #cycle;
end
endtask

ppi ppi_inst (.portA(tb_portA), .portB(tb_portB),
.portC(tb_portC), .rdb(tb_rdb),
.wrb(tb_wrb), .a2(tb_a2), .a1(tb_a1), .a0(tb_a0),
.reset(tb_reset),
.data(tb_data));

endmodule
```

Referring to Figure 6.35:

a. In the circle marked 1, ***reset*** occurs to reset the ***CWR*** register to "9e" and ***STATUS*** register to "0".
b. In the circle marked 2, ***wrb*** pulses low, causing a write to address "011" (***CWR*** register). The data on ***data[7:0]*** bus is "8e". This data is written into ***CWR*** register ("8e" in ***CWR*** register configures the PLB to function in mode 0 operation with ***portA***, ***portB***, and ***portC*** lower as input ports whereas ***portC*** upper is an output port).

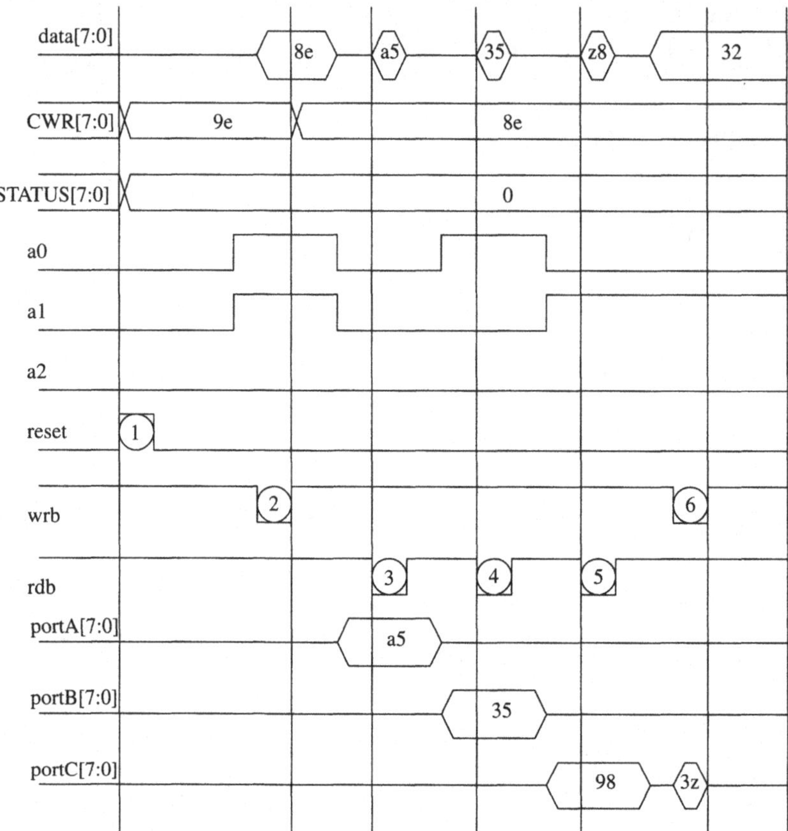

FIGURE 6.35. Diagram showing simulation results of test bench Example 6.3.

c. In the circle marked 3, ***rdb*** pulses low, causing a read from address "000". Because the PLB is operating in mode 0 with ***portA*** as input, the data at ***portA*** ("a5") is read to ***data[7:0]*** bus.

d. In the circle marked 4, ***rdb*** pulses low, causing a read from address "001". Because the PLB is operating in mode 0 with ***portB*** as input, the data at ***portB*** ("35") is read to ***data[7:0]*** bus.

e. In the circle marked 5, ***rdb*** pulses low, causing a read from address "010". Because the PLB is operating in mode 0 with ***portC*** lower as input and ***portC*** upper as output, the data at ***portA*** ("98") is read to ***data[7:0]*** bus as "z8" ("z" is for tri-state because portC upper is an output port).

f. In the circle marked 6, ***wrb*** pulses low, causing a write to address "010". Because the PLB is operating in mode 0 with ***portC*** lower as input and ***portC*** upper as output, the data at ***data[7:0]*** bus ("32") is written to ***portC*** as "3z" ("z" is for tri-state because portC lower is an input port).

6.6.3 Simulation for Mode 0 Operation with *PortA, PortB,* and *PortC* Upper as Input and *PortC* Lower as Output

In this simulation, ***portA***, ***portB***, and ***portC*** upper are configured as input ports whereas ***portC*** lower is an output port. The PLB operates in mode 0. Example 6.4 shows the Verilog code that can be used to simulate the PLB design for correct functionality.

Example 6.4 Simulation with *PortA, PortB,* and *PortC* Upper as Input and *PortC* Lower as Output in Mode 0 Operation

```
module ppi_tb ();

wire [7:0] tb_portA, tb_portB, tb_portC;
reg tb_rdb, tb_wrb, tb_reset;
wire [7:0] tb_data;
wire tb_a2, tb_a1, tb_a0;

reg [7:0] drive_portA, drive_portB, drive_portC,
drive_data, temp_data;

parameter cycle = 100;

assign tb_portA = drive_portA;
assign tb_portB = drive_portB;
assign tb_portC = drive_portC;
assign tb_data = drive_data;

reg [2:0] address;
assign tb_a2 = address [2];
assign tb_a1 = address [1];
assign tb_a0 = address [0];

initial
begin
     // for reset
     drive_portA = 8'hzz;
     drive_portB = 8'hzz;
     drive_portC = 8'hzz;
     tb_rdb = 1;
     tb_wrb = 1;
     address = 0;
     drive_data = 8'hzz;
     tb_reset = 0;

     #cycle;
```

```
     task_reset;

     // testing for mode 0 with all portA, portB
     // input, portCU
     // input, portCL output

     temp_data = 8'b10010110;
     CWR_write(temp_data);
     drive_data = 8'hzz;

     // read from portA
     address = 0;
     drive_portA = 8'ha5;
     read_port;
     drive_portA = 8'hzz;

     // read portB
     address = 1;
     drive_portB = 8'h35;
     read_port;
     drive_portB = 8'hzz;

     // read portC
     address = 2;
     drive_portC = 8'h98;
     read_port;
     drive_portC = 8'hzz;

     // write to portClower
     address = 2;
     drive_data = 8'h32;
     write_port;
     drive_data = 8'hzz;

end

task write_port;
begin
     tb_wrb = 1;
     tb_rdb = 1;
     #cycle;
     tb_wrb = 0;
     #cycle;
     tb_wrb = 1;
     #cycle;
end
endtask
```

```
task read_port;
begin
     tb_wrb = 1;
     tb_rdb = 1;
     #cycle;
     tb_rdb = 0;
     #cycle;
     tb_rdb = 1;
     #cycle;
end
endtask

task CWR_write;
input [7:0] temp_data;
begin
     tb_reset = 0;
     address = 3'b011;
     tb_rdb = 1;
     tb_wrb = 1;
     #cycle;
     drive_data = temp_data;
     tb_wrb = 0;
     #cycle;
     tb_wrb = 1;
     #cycle;
end
endtask

task task_reset;
begin
     tb_reset = 0;
     #cycle;
     tb_reset = 1;
     #cycle;
     tb_reset = 0;
     #cycle;
end
endtask

ppi ppi_inst (.portA(tb_portA), .portB(tb_portB),
.portC(tb_portC), .rdb(tb_rdb),
.wrb(tb_wrb), .a2(tb_a2), .a1(tb_a1), .a0(tb_a0),
.reset(tb_reset),
.data(tb_data));

endmodule
```

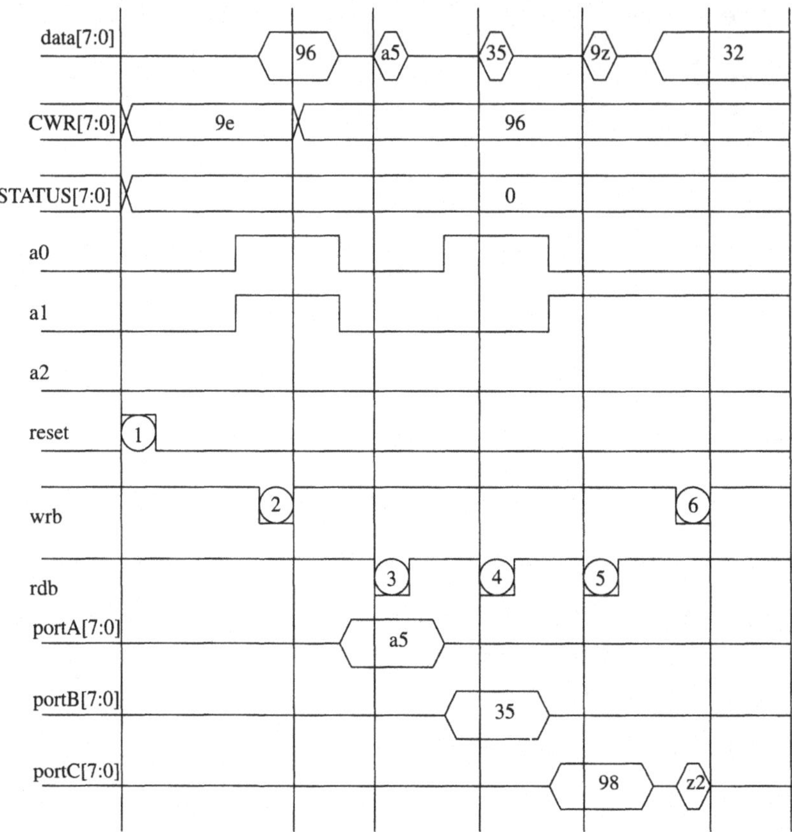

FIGURE 6.36. Diagram showing simulation results of test bench Example 6.4.

Referring to Figure 6.36:

a. In the circle marked 1, ***reset*** occurs to reset the ***CWR*** register to "9e" and the ***STATUS*** register to "0".

b. In the circle marked 2, ***wrb*** pulses low, causing a write to address "011" (***CWR*** register). The data on ***data[7:0]*** bus is "96". This data is written into ***CWR*** register ("96" in ***CWR*** register configures the PLB to function in mode 0 operation with ***portA***, ***portB***, and ***portC*** upper as input ports whereas ***portC*** lower is an output port).

c. In the circle marked 3, ***rdb*** pulses low, causing a read from address "000". Because the PLB is operating in mode 0 with ***portA*** as input, the data at ***portA*** ("a5") is read to ***data[7:0]*** bus.

d. In the circle marked 4, ***rdb*** pulses low, causing a read from address "001". Because the PLB is operating in mode 0 with ***portB*** as input, the data at ***portB*** ("35") is read to ***data[7:0]*** bus.

e. In the circle marked 5, ***rdb*** pulses low, causing a read from address "010". Because the PLB is operating in mode 0 with ***portC*** upper as input and ***portC*** lower as output, the data at ***portA*** ("98") is read to ***data[7:0]*** bus as "9z" ("z" is for tri-state because portC lower is an output port).

f. In the circle marked 6, ***wrb*** pulses low, causing a write to address "010". Because the PLB is operating in mode 0 with ***portC*** lower as output and ***portC*** upper as input, the data at ***data[7:0]*** bus ("32") is written to ***portC*** as "z2" ("z" is for tri-state because ***portC*** upper is an input port).

6.6.4 Simulation for Writing and Reading Data from STATUS and CWR Register

In this simulation, the functionality of writing and reading data from the ***CWR*** and ***STATUS*** registers are verified. Example 6.5 shows the Verilog code that can be used to simulate the PLB design for this functionality.

Example 6.5 Simulation for Writing and Reading Data from *STATUS* and *CWR* Register

```
module ppi_tb ();

wire [7:0] tb_portA, tb_portB, tb_portC;
reg tb_rdb, tb_wrb, tb_reset;
wire [7:0] tb_data;
wire tb_a2, tb_a1, tb_a0;

reg [7:0] drive_portA, drive_portB, drive_portC,
drive_data;

parameter cycle = 100;

assign tb_portA = drive_portA;
assign tb_portB = drive_portB;
assign tb_portC = drive_portC;
assign tb_data = drive_data;

reg [2:0] address;
assign tb_a2 = address [2];
assign tb_a1 = address [1];
assign tb_a0 = address [0];

initial
begin
     // for reset
     drive_portA = 8'hzz;
```

```
     drive_portB = 8'hzz;
     drive_portC = 8'hzz;
     tb_rdb = 1;
     tb_wrb = 1;
     address = 0;
     drive_data = 8'hzz;
     tb_reset = 1;

     #cycle;

     task_reset;

     // to write to STATUS
     address = 3'b111;
     drive_data = 8'b11111111;
     CWR_STATUS_write(address);
     drive_data = 8'hzz;

     // to write to CWR
     address = 3'b011;
     drive_data = 8'hba;
     CWR_STATUS_write(address);
     drive_data = 8'hzz;

     // read from STATUS reg
     address = 7;
     read_port;
     #cycle;

     // read from CWR reg
     address = 3;
     read_port;
     #cycle;

     #cycle;
end

task write_port;
begin
     tb_wrb = 1;
     tb_rdb = 1;
     #cycle;
     tb_wrb = 0;
     #cycle;
```

```
        tb_wrb = 1;
        #cycle;
end
endtask

task read_port;
begin
        tb_wrb = 1;
        tb_rdb = 1;
        #cycle;
        tb_rdb = 0;
        #cycle;
        tb_rdb = 1;
        #cycle;
end
endtask

task CWR_STATUS_write;
input [2:0] address;
begin
        tb_reset = 0;
        tb_rdb = 1;
        tb_wrb = 1;
        #cycle;
        tb_wrb = 0;
        #cycle;
        tb_wrb = 1;
        #cycle;
end
endtask

task task_reset;
begin
        tb_reset = 0;
        #cycle;
        tb_reset = 1;
        #cycle;
        tb_reset = 0;
        #cycle;
end
endtask

ppi ppi_inst (.portA(tb_portA), .portB(tb_portB),
.portC(tb_portC), .rdb(tb_rdb),
```

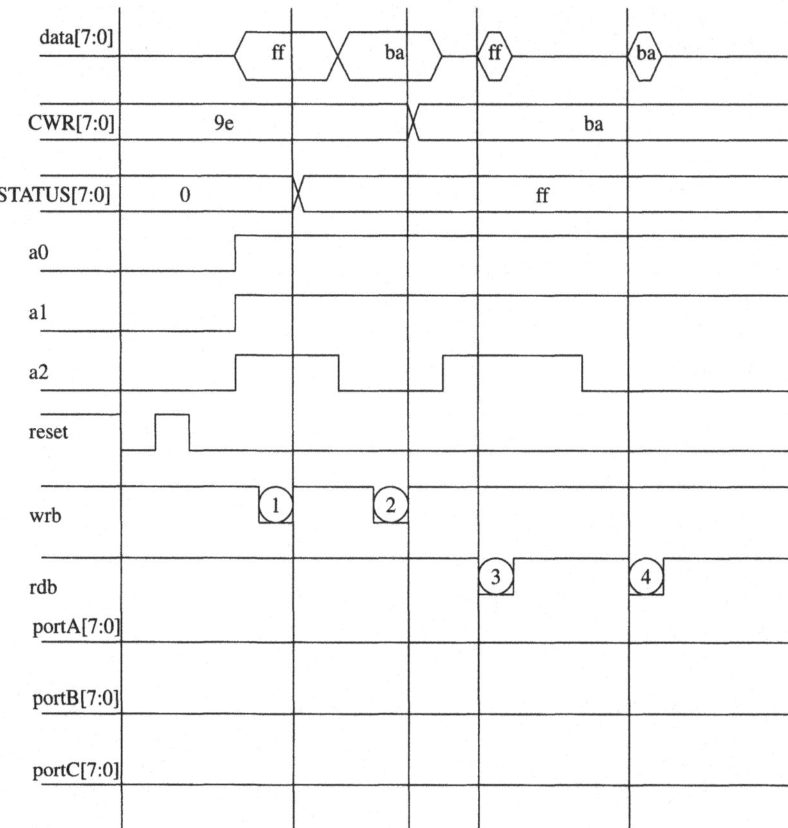

FIGURE 6.37. Diagram showing simulation results of test bench Example 6.5.

```
.wrb(tb_wrb), .a2(tb_a2), .a1(tb_a1), .a0(tb_a0),
.reset(tb_reset),
.data(tb_data));

endmodule
```

Referring to Figure 6.37:

a. In the circle marked 1, ***wrb*** pulses low, causing a write to address "111" (***STATUS*** register). The data on ***data[7:0]*** bus is "ff". This data is written into the ***STATUS*** register.
b. In the circle marked 2, ***wrb*** pulses low, causing a write to address "011" (***CWR*** register). The data on ***data[7:0]*** bus is "ba". This data is written into the ***CWR*** register.
c. In the circle marked 3, ***rdb*** pulses low, causing a read from address "111" (***STATUS*** register). The contents of the ***STATUS*** register are read onto ***data[7:0]*** bus ("ff").

d. In the circle marked 4, ***rdb*** pulses low, causing a read from address "011" (***CWR*** register). The contents of the ***CWR*** register are read onto ***data[7:0]*** bus ("ba").

6.6.5 Simulation for Mode 1 Operation with *PortA* and *PortB* as Strobed Input

In this simulation, ***portA*** and ***portB*** are configured as strobed input ports in mode 1 operation. In this mode, ***portC*** is used as control signals for the strobed input ports (***portA*** and ***portB***). Example 6.6 shows the Verilog code that can be used to simulate the PLB design for correct functionality.

Example 6.6 Simulation with *PortA* and *PortB* as Strobed Input in Mode 1 Operation

```
module ppi_tb ();

wire [7:0] tb_portA, tb_portB, tb_portC;
reg tb_rdb, tb_wrb, tb_reset;
wire [7:0] tb_data;
wire tb_a2, tb_a1, tb_a0;

reg [7:0] drive_portA, drive_portB, drive_portC,
drive_data;

parameter cycle = 100;

assign tb_portA = drive_portA;
assign tb_portB = drive_portB;
assign tb_portC = drive_portC;
assign tb_data = drive_data;

reg [2:0] address;
assign tb_a2 = address [2];
assign tb_a1 = address [1];
assign tb_a0 = address [0];

initial
begin
    // for reset
    drive_portA = 8'hzz;
    drive_portB = 8'hzz;
    drive_portC = 8'hzz;
    tb_rdb = 1;
    tb_wrb = 1;
```

```
    address = 0;
    drive_data = 8'hzz;
    tb_reset = 0;

    #cycle;

    task_reset;

    // for mode 1 with portA and portB input

    // to write to STATUS
    address = 3'b111;
    drive_data = 8'b11111111;
    CWR_STATUS_write(address);

    address = 3'b011;
    drive_data = 8'b10100110;
    // drive portC[4] to default 1
    drive_portC[4] = 1;
    // drive portC[2] to default 1
    drive_portC[2] = 1;
    CWR_STATUS_write(address);
    drive_data = 8'hzz;

    // read from portA
    address = 0;
    drive_portA = 8'ha5;
    drive_portB = 8'hba;
    drive_portC[4] = 0; // this is to have stbab at low
    drive_portC[2] = 1; // this is to have stbbb at high
    #cycle;
    drive_portC[4] = 1; // this is to have stbab back
    // at high
    read_port;
    address = 1;
    drive_portC[2] = 0; // this is to have stbbb at low
    #cycle;
    drive_portC[2] = 1; // this is to have stbbb back
    // at high
    #cycle;
    read_port;
    drive_portA = 8'hzz;
    drive_portB = 8'hzz;

end
```

```
task write_port;
begin
     tb_wrb = 1;
     tb_rdb = 1;
     #cycle;
     tb_wrb = 0;
     #cycle;
     tb_wrb = 1;
     #cycle;
end
endtask

task read_port;
begin
     tb_wrb = 1;
     tb_rdb = 1;
     #cycle;
     tb_rdb = 0;
     #cycle;
     tb_rdb = 1;
     #cycle;
end
endtask

task CWR_STATUS_write;
input [2:0] address;
begin
     tb_reset = 0;
     tb_rdb = 1;
     tb_wrb = 1;
     #cycle;
     tb_wrb = 0;
     #cycle;
     tb_wrb = 1;
     #cycle;
end
endtask

task task_reset;
begin
     tb_reset = 0;
     #cycle;
     tb_reset = 1;
     #cycle;
     tb_reset = 0;
     #cycle;
```

```
end
endtask

ppi ppi_inst (.portA(tb_portA), .portB(tb_portB),
.portC(tb_portC), .rdb(tb_rdb),
.wrb(tb_wrb), .a2(tb_a2), .a1(tb_a1), .a0(tb_a0),
.reset(tb_reset),
.data(tb_data));

endmodule
```

Referring to Figure 6.38:

a. In the circle marked 1, ***reset*** occurs to reset the ***CWR*** register to "9e" and the ***STATUS*** register to "0".

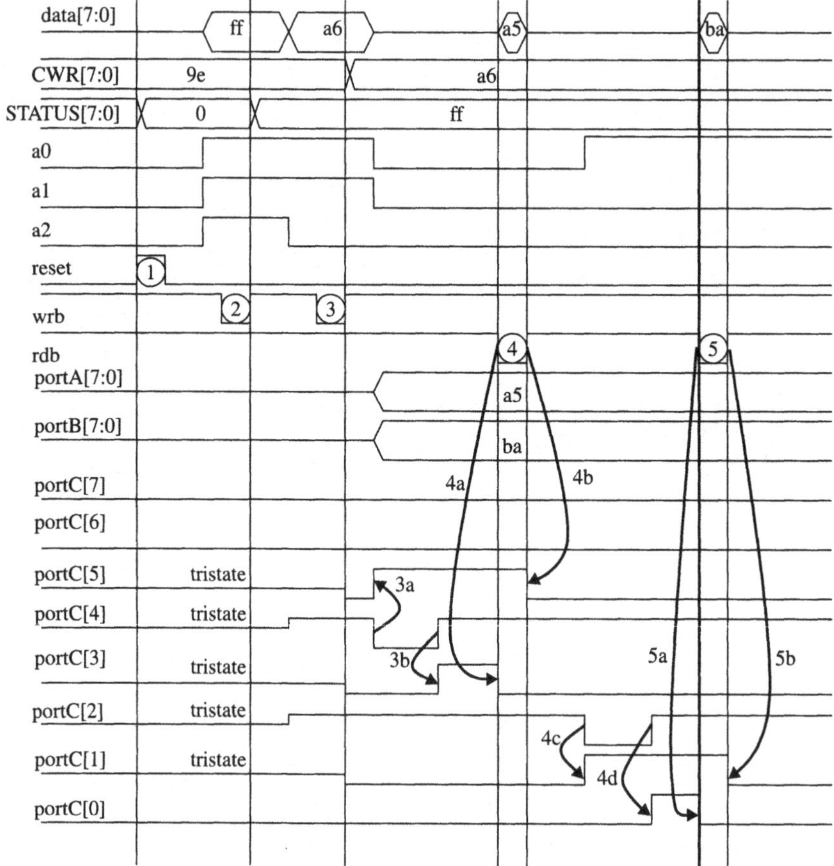

FIGURE 6.38. Diagram showing simulation results of test bench Example 6.6.

b. In the circle marked 2, ***wrb*** pulses low, causing a write to address "111" (***STATUS*** register). The data on ***data[7:0]*** bus is "ff". This data is written into the ***STATUS*** register. The contents of the ***STATUS*** register are used as a qualifier in generating the control signals ***portC*** when the PLB operates in mode 1 or mode 2 operation.
c. In the circle marked 3, ***wrb*** pulses low, causing a write to address "011" (***CWR*** register). The data on ***data[7:0]*** bus is "a6". This data is written into the ***CWR*** register. ("a6" in the ***CWR*** register configures the PLB to function in mode 1 operation with ***portA*** and ***portB*** as strobed input ports whereas ***portC*** acts as the control signals for ***portA*** and ***portB***.)
d. In the curved arrow marked 3a, ***portC[4]*** (input control signal ***stbab***) drives logic "0". This causes ***portC[5]*** (output control signal ***ibfa***) to be at logic "1".
e. In the curved arrow marked 3b, ***portC[4]*** (input control signal ***stbab***) drives logic "1". This causes ***portC[3]*** (output control signal ***intra***) to be at logic "1".
f. In the circle marked 4, ***rdb*** pulses low. The falling edge of ***rdb*** causes ***portC[3]*** (output control signal ***intra***) to be at logic "0". This is represented by curved arrow 4a.
g. The rising edge of ***rdb*** in the circle marked 4 causes ***portC[5]*** (output control signal ***ibfa***) to be at logic "0". This is represented by curved arrow 4b.
h. In the circle marked 4, when ***rdb*** pulses low, the data ("a5") at ***portA[7:0]*** is read onto the ***data[7:0]*** bus ("a5").
i. In the curved arrow marked 4c, ***portC[2]*** (input control signal ***stbbb***) drives logic "0". This causes ***portC[1]*** (output control signal ***ibfb***) to be at logic "1".
j. In the curved arrow marked 4d, ***portC[2]*** (input control signal ***stbbb***) drives logic "1". This causes ***portC[0]*** (output control signal ***intrb***) to be at logic "1".
k. In the circle marked 5, ***rdb*** pulses low. The falling edge of ***rdb*** causes ***portC[0]*** (output control signal ***intrb***) to be at logic "0". This is represented by curved arrow 5a.
l. The rising edge of ***rdb*** in the circle marked 5 causes ***portC[1]*** (output control signal ***ibfb***) to be at logic "0". This is represented by curved arrow 5b.
m. In the circle marked 5, when ***rdb*** pulses low, the data ("ba") at ***portB[7:0]*** is read onto the ***data[7:0]*** bus ("ba").

6.6.6 Simulation for Mode 1 Operation with *PortA* as Strobed Input and *PortB* as Strobed Output

In this simulation, ***portA*** is configured as a strobed input port and ***portB*** as a strobed output port in mode 1 operation. In this mode, ***portC*** is used as control signal for strobed input and output ports. Example 6.7 shows the Verilog code that can be used to simulate the PLB design for correct functionality.

Example 6.7 Simulation with *PortA* as Strobed Input and *PortB* as Strobed Output in Mode 1 Operation

```
module ppi_tb ();

wire [7:0] tb_portA, tb_portB, tb_portC;
reg tb_rdb, tb_wrb, tb_reset;
wire [7:0] tb_data;
wire tb_a2, tb_a1, tb_a0;

reg [7:0] drive_portA, drive_portB, drive_portC,
drive_data;

parameter cycle = 100;

assign tb_portA = drive_portA;
assign tb_portB = drive_portB;
assign tb_portC = drive_portC;
assign tb_data = drive_data;

reg [2:0] address;
assign tb_a2 = address [2];
assign tb_a1 = address [1];
assign tb_a0 = address [0];

initial
begin
     // for reset
     drive_portA = 8'hzz;
     drive_portB = 8'hzz;
     drive_portC = 8'hzz;
     tb_rdb = 1;
     tb_wrb = 1;
     address = 0;
     drive_data = 8'hzz;
     tb_reset = 0;

     #cycle;

     task_reset;

     // for mode 1 with portA input and portB output
```

```
    // to write to STATUS
    address = 3'b111;
    drive_data = 8'b11111111;
    CWR_STATUS_write(address);

    address = 3'b011;
    drive_data = 8'b10100010;
    // drive portC[4] to default 1 - this is stbab
    drive_portC[4] = 1;
    // drive portC[2] to default 1 - this is ackbb
    drive_portC[2] = 1;
    CWR_STATUS_write(address);
    drive_data = 8'hzz;

    // read from portA
    address = 0;
    drive_portA = 8'ha5;
    drive_portC[4] = 0; // this is to have stbab at low
    #cycle;
    drive_portC[4] = 1; // this is to have stbab back at high
    read_port;
    #cycle;
    drive_portA = 8'hzz;
    #cycle;

    // write to portB
    address = 1;
    drive_data = 8'hac;
    write_port;
    #cycle;
    drive_portC[2] = 0; // this is to have ackbb at low
    #cycle;
    drive_portC[2] = 1; // this is to have ackbb at high
    #cycle;

    #cycle;
end

task write_port;
begin
    tb_wrb = 1;
    tb_rdb = 1;
    #cycle;
    tb_wrb = 0;
    #cycle;
```

```
     tb_wrb = 1;
     #cycle;
end
endtask

task read_port;
begin
     tb_wrb = 1;
     tb_rdb = 1;
     #cycle;
     tb_rdb = 0;
     #cycle;
     tb_rdb = 1;
     #cycle;
end
endtask

task CWR_STATUS_write;
input [2:0] address;
begin
     tb_reset = 0;
     tb_rdb = 1;
     tb_wrb = 1;
     #cycle;
     tb_wrb = 0;
     #cycle;
     tb_wrb = 1;
     #cycle;
end
endtask

task task_reset;
begin
     tb_reset = 0;
     #cycle;
     tb_reset = 1;
     #cycle;
     tb_reset = 0;
     #cycle;
end
endtask

ppi ppi_inst (.portA(tb_portA), .portB(tb_portB),
.portC(tb_portC), .rdb(tb_rdb), .wrb(tb_wrb),
```

```
.a2(tb_a2), .a1(tb_a1), .a0(tb_a0), .reset(tb_reset),
.data(tb_data));

endmodule
```

Referring to Figure 6.39:

a. In the circle marked 1, ***reset*** occurs to reset the ***CWR*** register to "9e" and the ***STATUS*** register to "0".
b. In the circle marked 2, ***wrb*** pulses low, causing a write to address "111" (***STATUS*** register). The data on ***data[7:0]*** bus is "ff". This data is written into the ***STATUS*** register. The contents of the ***STATUS*** register are used as qualifiers in generating the control signals ***portC*** when the PLB operates in mode 1 or mode 2 operation.

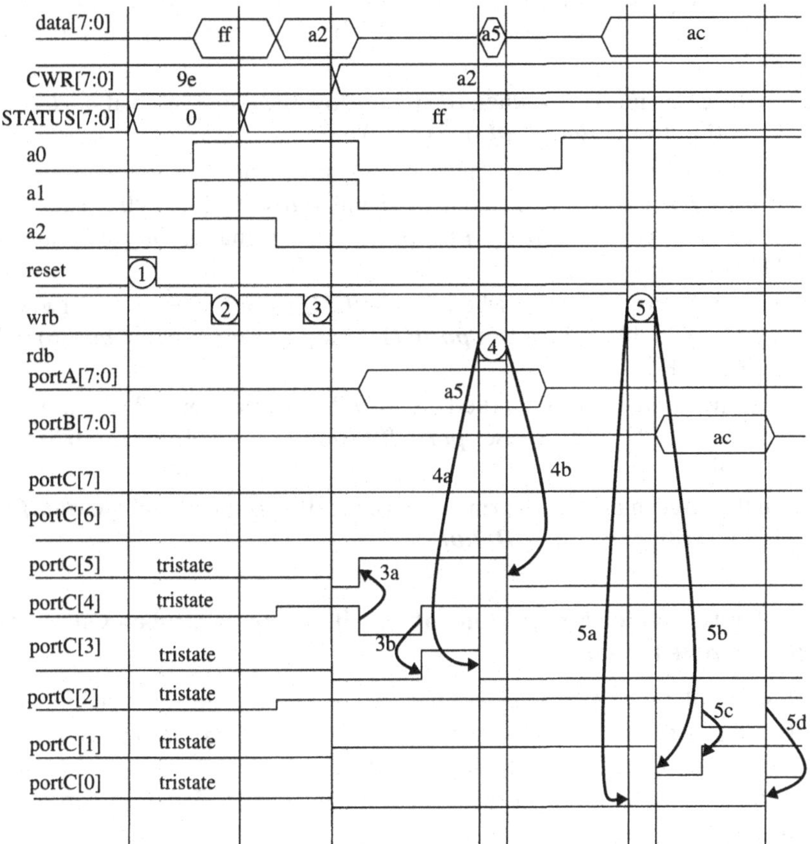

FIGURE 6.39. Diagram showing simulation results of test bench Example 6.7.

c. In the circle marked 3, ***wrb*** pulses low, causing a write to address "011" (***CWR*** register). The data on ***data[7:0]*** bus is "a2". This data is written into ***CWR*** register. ("a2" in ***CWR*** register configures the PLB to function in mode 1 operation with ***portA*** as a strobed input port and ***portB*** as a strobed output port. ***PortC*** acts as the control signals for the ***portA*** and ***portB***.)

d. For the curved arrow marked 3a, ***portC[4]*** (input control signal ***stbab***) drives logic "0". This causes ***portC[5]*** (output control signal ***ibfa***) to be at logic "1".

e. For the curved arrow marked 3b, ***portC[4]*** (input control signal ***stbab***) drives logic "1". This causes ***portC[3]*** (output control signal ***intra***) to be at logic "1".

f. For the circle marked 4, ***rdb*** pulses low. The falling edge of ***rdb*** causes ***portC[3]*** (output control signal ***intra***) to logic "0". This is represented by curved arrow 4a.

g. The rising edge of ***rdb*** in circle marked 4 causes ***portC[5]*** (output control signal ***ibfa***) to logic "0". This is represented by curved arrow 4b.

h. In the circle marked 4, when ***rdb*** pulses low, the data ("a5") at ***portA[7:0]*** is read onto the ***data[7:0]*** bus ("a5").

i. In the circle marked 5, ***wrb*** pulses low. The falling edge of ***wrb*** causes ***portC[0]*** (output control signal ***intrb***) to logic "0". The rising edge of ***wrb*** causes ***portC[1]*** (output control signal ***obfbb***) to be at logic "0".

j. For the curved arrow marked 5c, ***portC[2]*** (input control signal ***ackbb***) drives logic "0". This causes ***portC[1]*** (output control signal ***obfbb***) to be at logic "1".

k. For the curved arrow marked 5d, ***portC[2]*** (input control signal ***ackbb***) drives logic "1". This causes ***portC[0]*** (output control signal ***intrb***) to be at logic "1".

l. In the circle marked 5, during rising edge of wrb, data ("ac") at ***data[7:0]*** bus is written onto ***portB[7:0]***.

6.6.7 Simulation for Mode 1 Operation with *PortA* as Strobed Output and *PortB* as Strobed Input

In this simulation, ***portA*** is configured as a strobed output port and ***portB*** as a strobed input port in mode 1 operation. ***PortC*** is used as control signals for the strobed input and output ports. Example 6.8 shows the Verilog code that can be used to simulate the PLB design for correct functionality.

Example 6.8 Simulation with *PortA* as Strobed Output and *PortB* as Strobed Input in Mode 1 Operation

```
module ppi_tb ();

wire [7:0] tb_portA, tb_portB, tb_portC;
reg tb_rdb, tb_wrb, tb_reset;
wire [7:0] tb_data;
wire tb_a2, tb_a1, tb_a0;

reg [7:0] drive_portA, drive_portB, drive_portC,
drive_data;

parameter cycle = 100;

assign tb_portA = drive_portA;
assign tb_portB = drive_portB;
assign tb_portC = drive_portC;
assign tb_data = drive_data;

reg [2:0] address;
assign tb_a2 = address [2];
assign tb_a1 = address [1];
assign tb_a0 = address [0];

initial
begin
     // for reset
     drive_portA = 8'hzz;
     drive_portB = 8'hzz;
     drive_portC = 8'hzz;
     tb_rdb = 1;
     tb_wrb = 1;
     address = 0;
     drive_data = 8'hzz;
     tb_reset = 1;

     #cycle;

     task_reset;

     // for mode 1 with portA output and portB input

     // to write to STATUS
     address = 3'b111;
```

```
    drive_data = 8'b11111111;
    CWR_STATUS_write(address);

    address = 3'b011;
    drive_data = 8'b10100100;
    // drive portC[2] to default 1 - this is stbbb
    drive_portC[2] = 1;
    // drive portC[6] to default 1 - this is ackab
    drive_portC[6] = 1;
    CWR_STATUS_write(address);
    drive_data = 8'hzz;

    // write to portA
    address = 0;
    drive_data = 8'hac;
    write_port;
    #cycle;
    drive_portC[6] = 0; // this is to have ackab at low
    #cycle;
    drive_portC[6] = 1; // this is to have ackab at high
    #cycle;
    drive_data = 8'hzz;

    // read from portB
    address = 1;
    drive_portB = 8'ha5;
    #cycle;
    drive_portC[2] = 0; // this is to have stbbb at low
    #cycle;
    drive_portC[2] = 1; // this is to have stbbb back at high
    read_port;
    #cycle;
    drive_portB = 8'hzz;
    #cycle;

    #cycle;
end

task write_port;
begin
    tb_wrb = 1;
    tb_rdb = 1;
    #cycle;
    tb_wrb = 0;
    #cycle;
```

```
     tb_wrb = 1;
     #cycle;
end
endtask

task read_port;
begin
     tb_wrb = 1;
     tb_rdb = 1;
     #cycle;
     tb_rdb = 0;
     #cycle;
     tb_rdb = 1;
     #cycle;
end
endtask

task CWR_STATUS_write;
input [2:0] address;
begin
     tb_reset = 0;
     tb_rdb = 1;
     tb_wrb = 1;
     #cycle;
     tb_wrb = 0;
     #cycle;
     tb_wrb = 1;
     #cycle;
end
endtask

task task_reset;
begin
     tb_reset = 0;
     #cycle;
     tb_reset = 1;
     #cycle;
     tb_reset = 0;
     #cycle;
end
endtask

ppi ppi_inst (.portA(tb_portA), .portB(tb_portB),
.portC(tb_portC), .rdb(tb_rdb), .wrb(tb_wrb),
```

```
.a2(tb_a2), .a1(tb_a1), .a0(tb_a0), .reset(tb_reset),
.data(tb_data));

endmodule
```

Referring to Figure 6.40:

a. In the circle marked 1, ***reset*** occurs to reset the ***CWR*** register to "9e" and the ***STATUS*** register to "0".
b. In the circle marked 2, ***wrb*** pulses low, causing a write to address "111" (***STATUS*** register). The data on ***data[7:0]*** bus is "ff". This data is written into the ***STATUS*** register. The contents of the ***STATUS*** register are used as a qualifier in generating the control signals ***portC*** when the PLB operates in mode 1 or mode 2 operation.
c. In the circle marked 3, ***wrb*** pulses low, causing a write to address "011" (***CWR*** register). The data on ***data[7:0]*** bus is "a4". This data is written

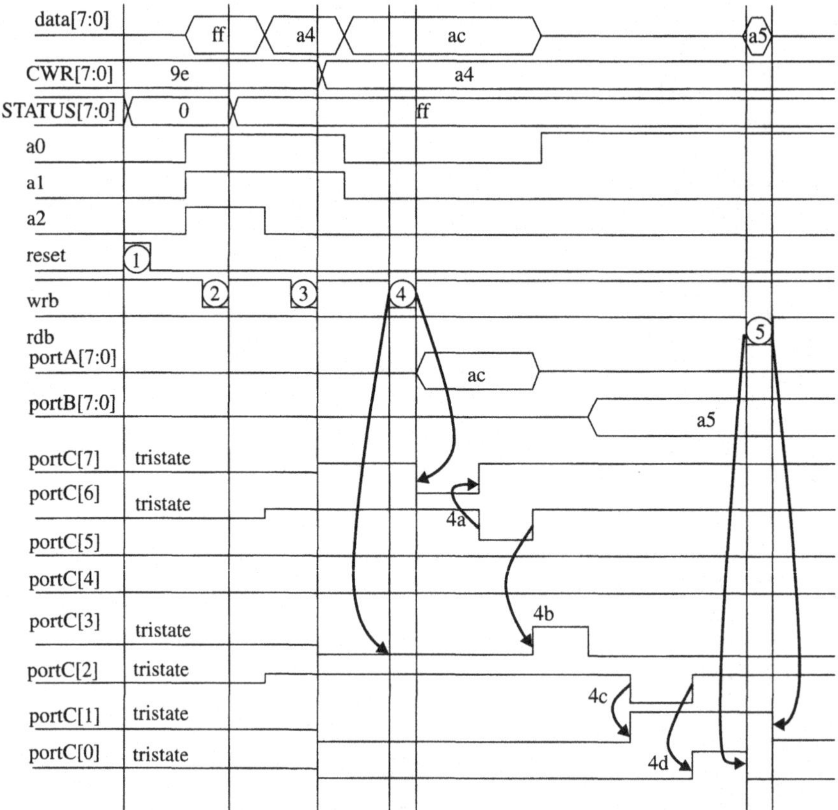

FIGURE 6.40. Diagram showing simulation results of test bench Example 6.8.

into the ***CWR*** register. ("a4" in ***CWR*** register configures the PLB to function in mode 1 operation with ***portA*** as a strobed output port and ***portB*** as a strobed input port. ***PortC*** acts as the control signals for ***portA*** and ***portB***.)

d. In the circle marked 4, ***wrb*** pulses low. The falling edge of ***wrb*** causes ***portC[3]*** (output control signal ***intra***) to be at logic "0". The rising edge of ***wrb*** in the circle marked 4 causes ***portC[7]*** (output control signal ***obfab***) to be at logic "0".

e. In the circle marked 4, during rising edge of ***wrb***, data ("ac") at ***data[7:0]*** bus is written onto ***portA[7:0]***.

f. For the curved arrow marked 4a, ***portC[6]*** (input control signal ***ackab***) drives logic "0". This causes ***portC[7]*** (output control signal ***obfab***) to be at logic "1".

g. For the curved arrow marked 4b, ***portC[6]*** (input control signal ***ackab***) drives logic "1". This causes ***portC[3]*** (output control signal ***intra***) to be at logic "1".

h. For the curved arrow marked 4c, ***portC[2]*** (input control signal ***stbbb***) drives logic "0". This causes ***portC[1]*** (output control signal ***ibfb***) to be at logic "1".

i. In the circle marked 5, ***rdb*** pulses low. The falling edge of ***rdb*** causes ***portC[0]*** (output control signal ***intrb***) to be at logic "0". The rising edge of ***rdb*** in the circle marked 5 causes ***portC[1]*** (output control signal ***ibfb***) to be at logic "0".

j. In the circle marked 4, when ***rdb*** pulses low, the data ("a5") at ***portB[7:0]*** is read onto ***data[7:0]*** bus ("a5").

6.6.8 Simulation for Mode 1 Operation with *PortA* and *PortB* as Strobed Output

In this simulation, ***portA*** and ***portB*** are configured as strobed output in mode 1 operation. ***PortC*** is used as control signals for the strobed input and output ports. Example 6.9 shows the Verilog code that can be used to simulate the PLB design for correct functionality.

Example 6.9 Simulation with *PortA* and *PortB* as Strobed Output in Mode 1 Operation

```
module ppi_tb ();

wire [7:0] tb_portA, tb_portB, tb_portC;
reg tb_rdb, tb_wrb, tb_reset;
wire [7:0] tb_data;
wire tb_a2, tb_a1, tb_a0;
```

```
reg [7:0] drive_portA, drive_portB, drive_portC,
drive_data;

parameter cycle = 100;

assign tb_portA = drive_portA;
assign tb_portB = drive_portB;
assign tb_portC = drive_portC;
assign tb_data = drive_data;

reg [2:0] address;
assign tb_a2 = address [2];
assign tb_a1 = address [1];
assign tb_a0 = address [0];

initial
begin
     // for reset
     drive_portA = 8'hzz;
     drive_portB = 8'hzz;
     drive_portC = 8'bz1z1z1zz;
     tb_rdb = 8'hzz;
     tb_wrb = 1;
     address = 0;
     drive_data = 8'hzz;
     tb_reset = 0;

     #cycle;

     task_reset;

     // for mode 1 with portA and portB output

     // to write to STATUS
     address = 3'b111;
     drive_data = 8'b11111111;
     CWR_STATUS_write(address);

     address = 3'b011;
     drive_data = 8'b10100000;
     CWR_STATUS_write(address);
     drive_data = 8'hzz;

     // write to portA
```

```
    // drive portC[6] to default 1 - this is ackab
    drive_portC[6] = 1;
    // drive portC[2] to default 1 - this is ackbb
    drive_portC[2] = 1;

    address = 0;
    drive_data = 8'ha5;
    write_port;
    #cycle;
    drive_portC[6] = 0; // this is to have ackab at low
    #cycle;
    drive_portC[6] = 1; // this is to have ackab back at high
    drive_portA = 8'hzz;
    #cycle;

    // write to portB
    address = 1;
    drive_data = 8'hac;
    write_port;
    #cycle;
    drive_portC[2] = 0; // this is to have ackbb at low
    #cycle;
    drive_portC[2] = 1; // this is to have ackbb at high
    drive_portB = 8'hzz;
    repeat (2) #cycle;
end

task write_port;
begin
    tb_wrb = 1;
    tb_rdb = 1;
    #cycle;
    tb_wrb = 0;
    #cycle;
    tb_wrb = 1;
    #cycle;
end
endtask

task read_port;
begin
    tb_wrb = 1;
    tb_rdb = 1;
    #cycle;
```

```
     tb_rdb = 0;
     #cycle;
     tb_rdb = 1;
     #cycle;
end
endtask

task CWR_STATUS_write;
input [2:0] address;
begin
     tb_reset = 0;
     tb_rdb = 1;
     tb_wrb = 1;
     #cycle;
     tb_wrb = 0;
     #cycle;
     tb_wrb = 1;
     #cycle;
end
endtask

task task_reset;
begin
     tb_reset = 0;
     #cycle;
     tb_reset = 1;
     #cycle;
     tb_reset = 0;
     #cycle;
end
endtask

ppi ppi_inst (.portA(tb_portA), .portB(tb_portB),
.portC(tb_portC), .rdb(tb_rdb), .wrb(tb_wrb),
.a2(tb_a2), .a1(tb_a1), .a0(tb_a0), .reset(tb_reset),
.data(tb_data));

endmodule
```

Referring to Figure 6.41:

a. In the circle marked 1, ***reset*** occurs to reset the ***CWR*** register to "9e" and the ***STATUS*** register to "0".
b. In the circle marked 2, ***wrb*** pulses low, causing a write to address "111" (***STATUS*** register). The data on ***data[7:0]*** bus is "ff". This data is written

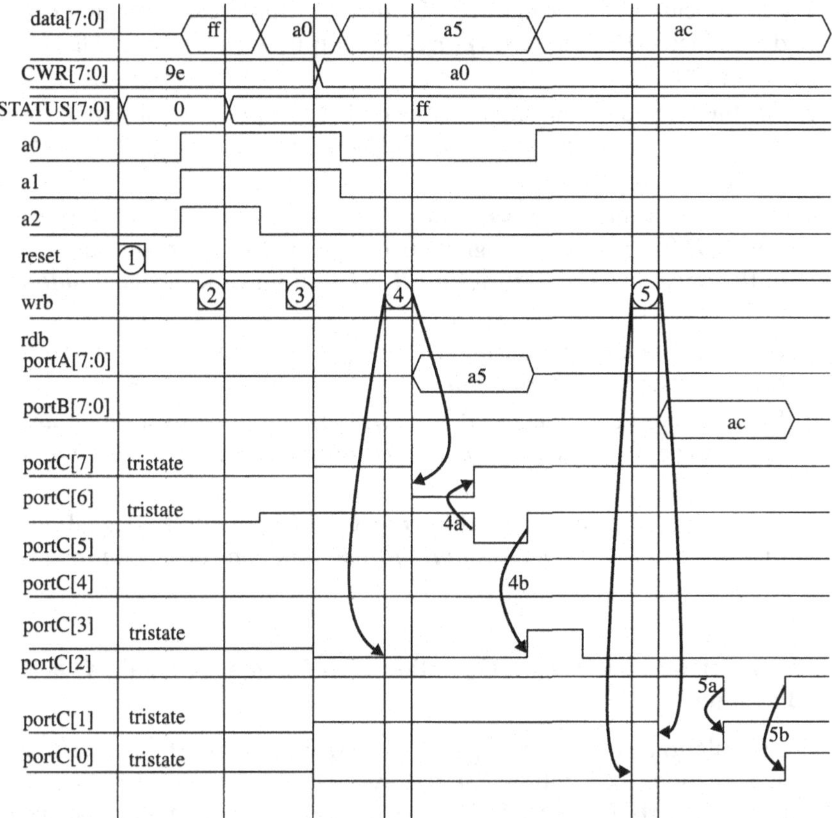

FIGURE 6.41. Diagram showing simulation results of test bench Example 6.9.

into the ***STATUS*** register. The contents of ***STATUS*** registers are used as a qualifier in generating the control signals ***portC*** when the PLB operates in mode 1 or mode 2 operation.

c. In the circle marked 3, ***wrb*** pulses low, causing a write to address "011" (***CWR*** register). The data on ***data[7:0]*** bus is "a0". This data is written into the ***CWR*** register. ("a0" in ***CWR*** register configures the PLB to function in mode 1 operation with ***portA*** and ***portB*** as strobed output. ***PortC*** acts as the control signals for ***portA*** and ***portB***.)

d. In the circle marked 4, ***wrb*** pulses low. The falling edge of ***wrb*** causes ***portC[3]*** (output control signal ***intra***) to logic "0". The rising edge of ***wrb*** in the circle marked 4 causes ***portC[7]*** (output control signal ***obfab***) to be at logic "0".

e. In the circle marked 4, during rising edge of ***wrb***, data ("a5") at ***data[7:0]*** bus is written onto ***portA[7:0]***.

f. For the curved arrow marked 4a, ***portC[6]*** (input control signal ***ackab***) drives logic "0". This causes ***portC[7]*** (output control signal ***obfab***) to be at logic "1".

g. For the curved arrow marked 4b, ***portC[6]*** (input control signal ***ackab***) drives logic "1". This causes ***portC[3]*** (output control signal ***intra***) to be at logic "1".

h. In the circle marked 5, ***wrb*** pulses low. The falling edge of ***wrb*** causes ***portC[0]*** (output control signal ***intrb***) to logic "0". The rising edge of ***wrb*** in the circle marked 5 causes ***portC[1]*** (output control signal ***obfbb***) to logic "0".

i. In the circle marked 5, during rising edge of ***wrb***, data ("ac") at ***data[7:0]*** bus is written onto ***portB[7:0]***.

j. In the curved arrow marked 5a, ***portC[2]*** (input control signal ***ackbb***) drives logic "0". This causes ***portC[1]*** (output control signal ***obfbb***) to be at logic "1".

k. In the curved arrow marked 5b, ***portC[2]*** (input control signal ***ackbb***) drives logic "1". This causes ***portC[0]*** (output control signal ***intrb***) to be at logic "1".

6.6.9 Simulation for Mode 2 Operation with *PortA* as Strobed I/O and PortB as Input

In this simulation, ***portA*** is configured as a strobed I/O port and ***portB*** as an input port in mode 2 operation. ***PortC*** is used as control signals for the strobed I/O and input ports. Example 6.10 shows the Verilog code that can be used to simulate the PLB design for correct functionality.

Example 6.10 Simulation with *PortA* as Strobed I/O and *PortB* as Input in Mode 2 Operation

```
module ppi_tb ();

wire [7:0] tb_portA, tb_portB, tb_portC;
reg tb_rdb, tb_wrb, tb_reset;
wire [7:0] tb_data;
wire tb_a2, tb_a1, tb_a0;

reg [7:0] drive_portA, drive_portB, drive_portC,
drive_data;

parameter cycle = 100;

assign tb_portA = drive_portA;
assign tb_portB = drive_portB;
```

```
assign tb_portC = drive_portC;
assign tb_data = drive_data;

reg [2:0] address;
assign tb_a2 = address [2];
assign tb_a1 = address [1];
assign tb_a0 = address [0];

initial
begin
     // for reset
     drive_portA = 8'hzz;
     drive_portB = 8'hzz;
     drive_portC = 8'hzz;
     tb_rdb = 1;
     tb_wrb = 1;
     address = 0;
     drive_data = 8'hzz;
     tb_reset = 0;

     #cycle;

     task_reset;

     // for mode 2 with portB input

     // to write to STATUS
     address = 3'b111;
     drive_data = 8'b11111111;
     CWR_STATUS_write(address);

     address = 3'b011;
     drive_data = 8'b11000110;
     // drive portC[4] to default 1
     drive_portC[4] = 1;
     // drive portC[6] to default 1
     drive_portC[6] = 1;
     CWR_STATUS_write(address);
     drive_data = 8'hzz;

     // read from portA
     address = 0;
     drive_portA = 8'ha5;
     drive_portC[4] = 0; // this is to have stbab at low
     #cycle;
```

```
     drive_portC[4] = 1; // this is to have stbab back at high
     read_port;
     #cycle;
     drive_portA = 8'hzz;

     // write to portA

     drive_data = 8'haa;
     write_port;
     #cycle;
     drive_portC[6] = 0; // this is to have ackab at low
     #cycle;
     drive_portC[6] = 1; // this is to have ackab back at high
     drive_portA = 8'hzz;
     #cycle;
     drive_data = 8'hzz;
     #cycle;

     // read from portB
     address = 1;
     drive_portB = 8'h35;
     read_port;
     drive_portB = 8'hzz;

     #cycle;
end

task write_port;
begin
     tb_wrb = 1;
     tb_rdb = 1;
     #cycle;
     tb_wrb = 0;
     #cycle;
     tb_wrb = 1;
     #cycle;
end
endtask

task read_port;
begin
     tb_wrb = 1;
     tb_rdb = 1;
     #cycle;
     tb_rdb = 0;
```

```
        #cycle;
        tb_rdb = 1;
        #cycle;
end
endtask

task CWR_STATUS_write;
input [2:0] address;
begin
        tb_reset = 0;
        tb_rdb = 1;
        tb_wrb = 1;
        #cycle;
        tb_wrb = 0;
        #cycle;
        tb_wrb = 1;
        #cycle;
end
endtask

task task_reset;
begin
        tb_reset = 0;
        #cycle;
        tb_reset = 1;
        #cycle;
        tb_reset = 0;
        #cycle;
end
endtask

ppi ppi_inst (.portA(tb_portA), .portB(tb_portB),
.portC(tb_portC), .rdb(tb_rdb), .wrb(tb_wrb),
.a2(tb_a2), .a1(tb_a1), .a0(tb_a0), .reset(tb_reset),
.data(tb_data));

endmodule
```

Referring to Figure 6.42:

a. In the circle marked 1, ***reset*** occurs to reset the ***CWR*** register to "9e" and the ***STATUS*** register to "0".
b. In the circle marked 2, ***wrb*** pulses low, causing a write to address "111" (***STATUS*** register). The data on ***data[7:0]*** bus is "ff". This data is written into the ***STATUS*** register. The contents of the ***STATUS*** register are used

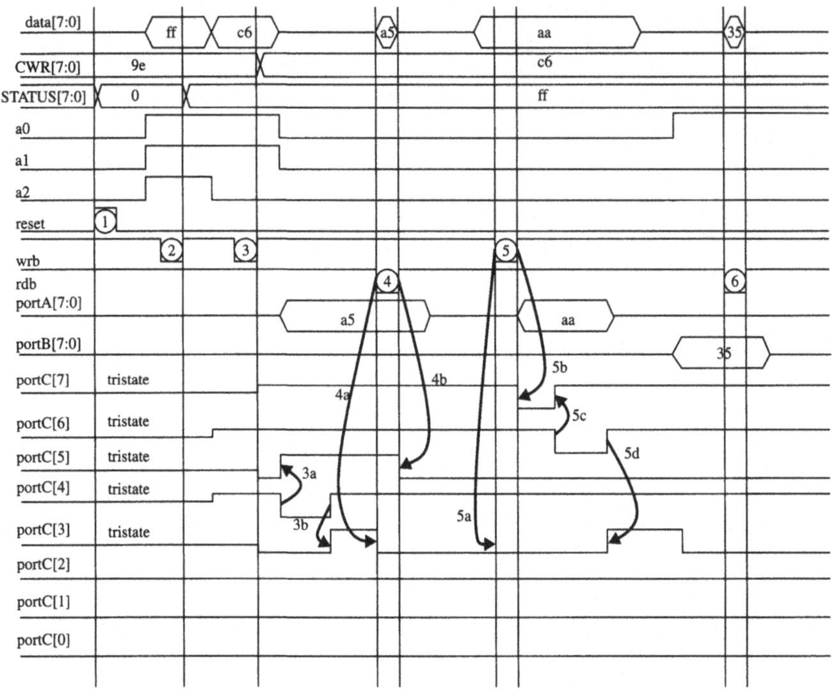

FIGURE 6.42. Diagram showing simulation results of test bench Example 6.10.

as a qualifier in generating the control signals ***portC*** when the PLB operates in mode 1 or mode 2 operation.

c. In the circle marked 3, ***wrb*** pulses low, causing a write to address "011" (***CWR*** register). The data on ***data[7:0]*** bus is "c6". This data is written into the ***CWR*** register. ("c6" in the ***CWR*** register configures the PLB to function in mode 2 operation with ***portA*** as a strobed I/O port and ***portB*** as an input port. ***PortC*** acts as the control signals for ***portA***.)

d. For the curved arrow marked 3a, ***portC[4]*** (input control signal ***stbab***) drives logic "0". This causes ***portC[5]*** (output control signal ***ibfa***) to be at logic "1".

e. For the curved arrow marked 3b, ***portC[4]*** (input control signal ***stbab***) drives logic "1". This causes ***portC[3]*** (output control signal ***intra***) to be at logic "1".

f. In the circle marked 4, ***rdb*** pulses low. The falling edge of ***rdb*** causes ***portC[3]*** (output control signal ***intra***) to logic "0". This is represented by curved arrow 4a.

g. The rising edge of ***rdb*** in the circle marked 4 causes ***portC[5]*** (output control signal ***ibfa***) to logic "0". This is represented by curved arrow 4b.

h. In the circle marked 4, when ***rdb*** pulses low, the data ("a5") at ***portA[7:0]*** is read onto the ***data[7:0]*** bus ("a5").

i. In the circle marked 5, ***wrb*** pulses low. The falling edge of ***wrb*** causes ***portC[3]*** (output control signal ***intra***) to be at logic "0". The rising edge of ***wrb*** in the circle marked 5 causes ***portC[7]*** (output control signal ***obfab***) to be at logic "0".

j. In the circle marked 5, during rising edge of ***wrb***, data ("aa") at ***data[7:0]*** bus is written onto ***portA[7:0]***.

k. For the curved arrow marked 5c, ***portC[6]*** (input control signal ***ackab***) drives logic "0". This causes ***portC[7]*** (output control signal ***obfab***) to be at logic "1".

l. For the curved arrow marked 5d, ***portC[6]*** (input control signal ***ackab***) drives logic "1". This causes ***portC[3]*** (output control signal ***intra***) to be at logic "1".

i. In the circle marked 6, ***rdb*** pulses low, causing a read from address "001". Because the PLB is operating in mode 2 with ***portB*** as input, the data at ***portB*** ("35") is read to ***data[7:0]*** bus.

6.6.10 Simulation for Mode 2 Operation with *PortA* as Strobed I/O and *PortB* as Output

In this simulation, ***portA*** is configured as a strobed I/O port and ***portB*** as an output port in mode 2 operation. ***PortC*** is used as control signals for the strobed I/O and output ports. Example 6.11 shows the Verilog code that can be used to simulate the PLB design for correct functionality.

Example 6.11 Simulation with *PortA* as Strobed I/O and *PortB* as Output in Mode 2 Operation

```
module ppi_tb ();

wire [7:0] tb_portA, tb_portB, tb_portC;
reg tb_rdb, tb_wrb, tb_reset;
wire [7:0] tb_data;
wire tb_a2, tb_a1, tb_a0;

reg [7:0] drive_portA, drive_portB, drive_portC,
drive_data;

parameter cycle = 100;

assign tb_portA = drive_portA;
assign tb_portB = drive_portB;
```

```
assign tb_portC = drive_portC;
assign tb_data = drive_data;

reg [2:0] address;
assign tb_a2 = address [2];
assign tb_a1 = address [1];
assign tb_a0 = address [0];

initial
begin
     // for reset
     drive_portA = 8'hzz;
     drive_portB = 8'hzz;
     drive_portC = 8'hzz;
     tb_rdb = 1;
     tb_wrb = 1;
     address = 0;
     drive_data = 8'hzz;
     tb_reset = 0;

     #cycle;

     task_reset;

     // for mode 2 with portB input

     // to write to STATUS
     address = 3'b111;
     drive_data = 8'b11111111;
     CWR_STATUS_write(address);

     address = 3'b011;
     drive_data = 8'b11000010;
     // drive portC[4] to default 1
     drive_portC[4] = 1;
     // drive portC[6] to default 1
     drive_portC[6] = 1;
     CWR_STATUS_write(address);
     drive_data = 8'hzz;

     // read from portA
     address = 0;
     drive_portA = 8'ha5;
     drive_portC[4] = 0; // this is to have stbab at low
     #cycle;
```

```
    drive_portC[4] = 1; // this is to have stbab back at high
    read_port;
    #cycle;
    drive_portA = 8'hzz;

    // write to portA

    drive_data = 8'haa;
    write_port;
    #cycle;
    drive_portC[6] = 0; // this is to have ackab at low
    #cycle;
    drive_portC[6] = 1; // this is to have ackab back at high
    drive_portA = 8'hzz;
    #cycle;
    drive_data = 8'hzz;
    #cycle;

    // write to portB
    address = 1;
    drive_data = 8'h67;
    write_port;
    drive_data = 8'hzz;
    #cycle;

end

task write_port;
begin
    tb_wrb = 1;
    tb_rdb = 1;
    #cycle;
    tb_wrb = 0;
    #cycle;
    tb_wrb = 1;
    #cycle;
end
endtask

task read_port;
begin
    tb_wrb = 1;
    tb_rdb = 1;
    #cycle;
    tb_rdb = 0;
```

```
        #cycle;
        tb_rdb = 1;
        #cycle;
end
endtask

task CWR_STATUS_write;
input [2:0] address;
begin
        tb_reset = 0;
        tb_rdb = 1;
        tb_wrb = 1;
        #cycle;
        tb_wrb = 0;
        #cycle;
        tb_wrb = 1;
        #cycle;
end
endtask

task task_reset;
begin
        tb_reset = 0;
        #cycle;
        tb_reset = 1;
        #cycle;
        tb_reset = 0;
        #cycle;
end
endtask

ppi ppi_inst (.portA(tb_portA), .portB(tb_portB),
.portC(tb_portC), .rdb(tb_rdb), .wrb(tb_wrb),
.a2(tb_a2), .a1(tb_a1), .a0(tb_a0), .reset(tb_reset),
.data(tb_data));

endmodule
```

Referring to Figure 6.43:

a. In the circle marked 1, ***reset*** occurs to reset the ***CWR*** register to "9e" and the ***STATUS*** register to "0".
b. In the circle marked 2, ***wrb*** pulses low, causing a write to address "111" (***STATUS*** register). The data on ***data[7:0]*** bus is "ff". This data is written into the ***STATUS*** register. The contents of the ***STATUS*** register are used

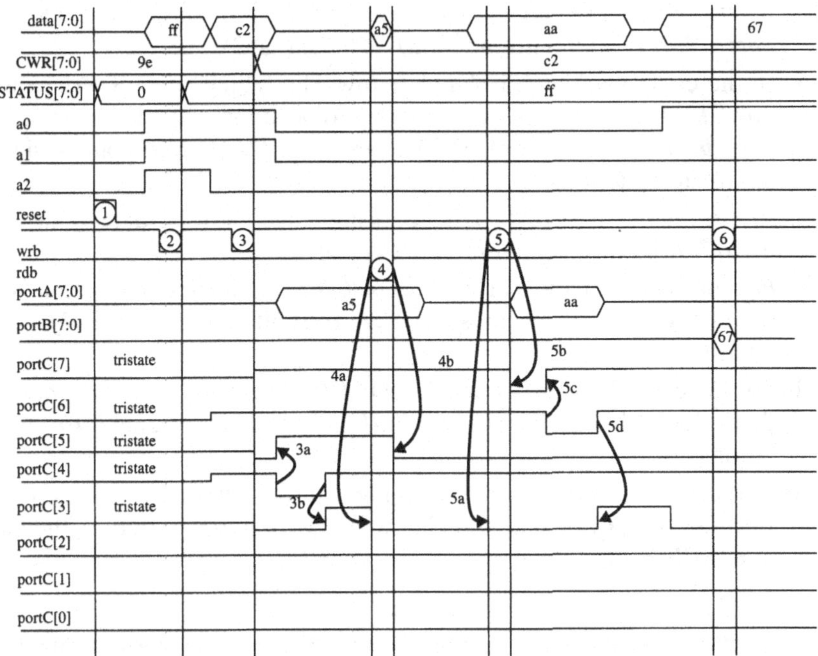

FIGURE 6.43. Diagram showing simulation results of test bench Example 6.11.

as a qualifier in generating the control signals ***portC*** when the PLB operates in mode 1 or mode 2 operation.

c. In the circle marked 3, ***wrb*** pulses low, causing a write to address "011" (***CWR*** register). The data on ***data[7:0]*** bus is "c2". This data is written into the ***CWR*** register. ("c2" in the ***CWR*** register configures the PLB to function in mode 2 operation with ***portA*** as a strobed I/O port and ***portB*** as an output port. ***PortC*** acts as the control signals for ***portA***.)

d. For the curved arrow marked 3a, ***portC[4]*** (input control signal ***stbab***) drives logic "0". This causes ***portC[5]*** (output control signal ***ibfa***) to be at logic "1".

e. For the curved arrow marked 3b, ***portC[4]*** (input control signal ***stbab***) drives logic "1". This causes ***portC[3]*** (output control signal ***intra***) to be at logic "1".

f. In the circle marked 4, ***rdb*** pulses low. The falling edge of ***rdb*** causes ***portC[3]*** (output control signal ***intra***) to logic "0". This is represented by curved arrow 4a.

g. The rising edge of ***rdb*** in the circle marked 4 causes ***portC[5]*** (output control signal ***ibfa***) to be at logic "0". This is represented by curved arrow 4b.

h. In the circle marked 4, when ***rdb*** pulses low, the data ("a5") at ***portA[7:0]*** is read onto the ***data[7:0]*** bus ("a5").

i. In the circle marked 5, ***wrb*** pulses low. The falling edge of ***wrb*** causes ***portC[3]*** (output control signal ***intra***) to be at logic "0". The rising edge of ***wrb*** in the circle marked 5 causes ***portC[7]*** (output control signal ***obfab***) to be at logic "0".

j. In the circle marked 5, during rising edge of ***wrb***, data ("aa") at ***data[7:0]*** bus is written onto ***portA[7:0]***.

k. For the curved arrow marked 5c, ***portC[6]*** (input control signal ***ackab***) drives logic "0". This causes ***portC[7]*** (output control signal ***obfab***) to be at logic "1".

l. For the curved arrow marked 5d, ***portC[6]*** (input control signal ***ackab***) drives logic "1". This causes ***portC[3]*** (output control signal ***intra***) to be at logic "1".

m. In the circle marked 6, ***wrb*** pulses low, causing a write to address "001". Because the PLB is operating in mode 2 with ***portB*** as output, the data at ***data[7:0]*** bus ("67") is written to ***portB***.

6.6.11 Simulation for Mode 1 Operation with *PortA* and *PortB* as Strobed Input and *STATUS* Register Disabled

In this simulation, ***portA*** and ***portB*** are configured as strobed input ports in mode 1 operation with ***STATUS*** register disabled (contents of ***STATUS*** register are "00000000"). In this mode, ***portC*** is used as control signals for the strobed input ports (***portA*** and ***portB***). Example 6.12 shows the Verilog code that can be used to simulate the PLB design for correct functionality.

Example 6.12 Simulation with *PortA* and *PortB* as Strobed Input in Mode 1 Operation with *STATUS* Register Disabled

```
module ppi_tb ();

wire [7:0] tb_portA, tb_portB, tb_portC;
reg tb_rdb, tb_wrb, tb_reset;
wire [7:0] tb_data;
wire tb_a2, tb_a1, tb_a0;

reg [7:0] drive_portA, drive_portB, drive_portC,
drive_data;

parameter cycle = 100;

assign tb_portA = drive_portA;
assign tb_portB = drive_portB;
```

```
assign tb_portC = drive_portC;
assign tb_data = drive_data;

reg [2:0] address;
assign tb_a2 = address [2];
assign tb_a1 = address [1];
assign tb_a0 = address [0];

initial
begin
    // for reset
    drive_portA = 8'hzz;
    drive_portB = 8'hzz;
    drive_portC = 8'hzz;
    tb_rdb = 1;
    tb_wrb = 1;
    address = 0;
    drive_data = 8'hzz;
    tb_reset = 0;

    #cycle;

    task_reset;

    // for mode 1 with portA and portB input

    // to write to STATUS
    address = 3'b111;
    drive_data = 8'h00;
    CWR_STATUS_write(address);

    address = 3'b011;
    drive_data = 8'b10100110;
    // drive portC[4] to default 1
    drive_portC[4] = 1;
    // drive portC[2] to default 1
    drive_portC[2] = 1;
    CWR_STATUS_write(address);
    drive_data = 8'hzz;

    // read from portA
    address = 0;
    drive_portA = 8'ha5;
    drive_portB = 8'hba;
    drive_portC[4] = 0; // this is to have stbab at low
```

```
        drive_portC[2] = 1; // this is to have stbbb at high
        #cycle;
        drive_portC[4] = 1; // this is to have stbab back at high
        read_port;
        address = 1;
        drive_portC[2] = 0; // this is to have stbbb at low
        #cycle;
        drive_portC[2] = 1; // this is to have stbbb back
        #cycle;
        read_port;
        drive_portA = 8'hzz;
        drive_portB = 8'hzz;


end


task write_port;
begin
        tb_wrb = 1;
        tb_rdb = 1;
        #cycle;
        tb_wrb = 0;
        #cycle;
        tb_wrb = 1;
        #cycle;
end
endtask

task read_port;
begin
        tb_wrb = 1;
        tb_rdb = 1;
        #cycle;
        tb_rdb = 0;
        #cycle;
        tb_rdb = 1;
        #cycle;
end
endtask

task CWR_STATUS_write;
input [2:0] address;
begin
        tb_reset = 0;
        tb_rdb = 1;
```

```
        tb_wrb = 1;
        #cycle;
        tb_wrb = 0;
        #cycle;
        tb_wrb = 1;
        #cycle;
end
endtask

task task_reset;
begin
        tb_reset = 0;
        #cycle;
        tb_reset = 1;
        #cycle;
        tb_reset = 0;
        #cycle;
end
endtask

ppi ppi_inst (.portA(tb_portA), .portB(tb_portB),
.portC(tb_portC), .rdb(tb_rdb), .wrb(tb_wrb),
.a2(tb_a2), .a1(tb_a1), .a0(tb_a0), .reset(tb_reset),
.data(tb_data));

endmodule
```

Referring to Figure 6.44:

a. In the circle marked 1, ***reset*** occurs to reset the ***CWR*** register to "9e" and the ***STATUS*** register to "0".
b. In the circle marked 2, ***wrb*** pulses low, causing a write to address "111" (***STATUS*** register). The data on ***data[7:0]*** bus is "0". This data is written into the ***STATUS*** register. The contents of the ***STATUS*** register are used as a qualifier in generating the control signals ***portC*** when the PLB operates in mode 1 or mode 2 operation.
c. In the circle marked 3, ***wrb*** pulses low, causing a write to address "011" (***CWR*** register). The data on ***data[7:0]*** bus is "a6". This data is written into the ***CWR*** register. ("a6" in the ***CWR*** register configures the PLB to function in mode 1 operation with ***portA*** and ***portB*** as strobed input ports while ***portC*** acts as the control signals for ***portA*** and ***portB***.)
d. For the curved arrow marked 3a, ***portC[4]*** (input control signal ***stbab***) drives logic "0". This causes ***portC[5]*** (output control signal ***ibfa***) to be at logic "1".

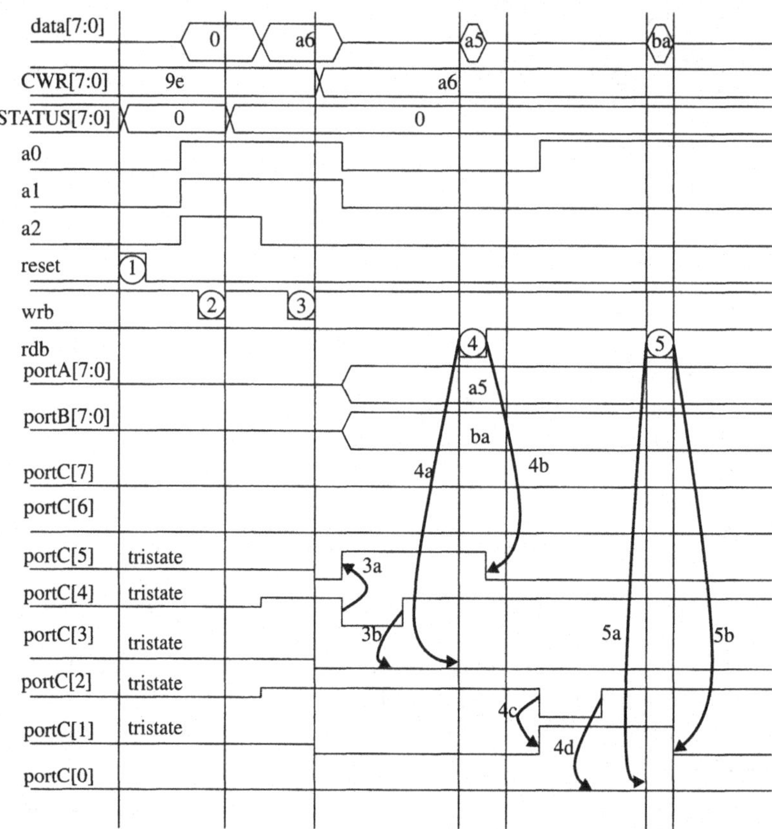

FIGURE 6.44. Diagram showing simulation results of test bench Example 6.12.

e. For the curved arrow marked 3b, ***portC[4]*** (input control signal ***stbab***) drives logic "1". This causes ***portC[3]*** (output control signal ***intra***) to be at logic "1". However, because the qualifier for ***portC[3]*** is ***STATUS[0]*** (content of ***STATUS[0]*** is "0"), ***portC[3]*** remains at logic "0" instead of toggling to logic "1".

f. For the circle marked 4, ***rdb*** pulses low. The falling edge of ***rdb*** causes ***portC[3]*** (output control signal ***intra***) to logic "0". This is represented by the curved arrow 4a. However, from Figure 6.44, ***portC[3]*** is already at logic "0" because the qualifier ***STATUS[0]*** is at logic "0".

g. The rising edge of ***rdb*** in the circle marked 4 causes ***portC[5]*** (output control signal ***ibfa***) to be at logic "0". This is represented by curved arrow 4b.

h. In the circle marked 4, when ***rdb*** pulses low, the data ("a5") at ***portA[7:0]*** is read onto the ***data[7:0]*** bus ("a5").

i. For the curved arrow marked 4c, ***portC[2]*** (input control signal ***stbbb***) drives logic "0". This causes ***portC[1]*** (output control signal ***ibfb***) to be at logic "1".

j. For the curved arrow marked 4d, ***portC[2]*** (input control signal ***stbbb***) drives logic "1". This causes ***portC[0]*** (output control signal ***intrb***) to be at logic "1". However, because the qualifier for ***portC[0]*** is ***STATUS[1]*** (content of ***STATUS[1]*** is "0"), ***portC[0]*** remains at logic "0" instead of toggling to logic "1".

k. In the circle marked 5, ***rdb*** pulses low. The falling edge of ***rdb*** causes ***portC[0]*** (output control signal ***intrb***) to be at logic "0". This is represented by curved arrow 5a. However from Figure 6.44, ***portC[0]*** is already at logic "0" because the qualifier ***STATUS[1]*** is at logic "0".

l. The rising edge of ***rdb*** in the circle marked 5 causes ***portC[1]*** (output control signal ***ibfb***) to be at logic "0". This is represented by curved arrow 5b.

m. In the circle marked 5, when ***rdb*** pulses low, the data ("ba") at ***portB[7:0]*** is read onto the ***data[7:0]*** bus ("ba").

6.6.12 Simulation for Mode 2 Operation with *PortA* as Strobed I/O and *PortB* as Output and *STATUS* Register Disabled

In this simulation, ***portA*** is configured as a strobed I/O port and ***portB*** as an output port in mode 2 operation with the ***STATUS*** register disabled (contents of the ***STATUS*** register are "00000000"). ***PortC*** is used as control signals for the strobed I/O and output port. Example 6.13 shows the Verilog code that can be used to simulate the PLB design for correct functionality.

Example 6.13 Simulation with *PortA* as Strobed I/O and *PortB* as Output in Mode 2 Operation with *STATUS* Register Disabled

```
module ppi_tb ();

wire [7:0] tb_portA, tb_portB, tb_portC;
reg tb_rdb, tb_wrb, tb_reset;
wire [7:0] tb_data;
wire tb_a2, tb_a1, tb_a0;

reg [7:0] drive_portA, drive_portB, drive_portC,
drive_data;

parameter cycle = 100;

assign tb_portA = drive_portA;
assign tb_portB = drive_portB;
```

```
assign tb_portC = drive_portC;
assign tb_data = drive_data;

reg [2:0] address;
assign tb_a2 = address [2];
assign tb_a1 = address [1];
assign tb_a0 = address [0];

initial
begin
     // for reset
     drive_portA = 8'hzz;
     drive_portB = 8'hzz;
     drive_portC = 8'hzz;
     tb_rdb = 1;
     tb_wrb = 1;
     address = 0;
     drive_data = 8'hzz;
     tb_reset = 0;

     #cycle;

     task_reset;

     // for mode 2 with portB input

     // to write to STATUS
     address = 3'b111;
     drive_data = 8'h00;
     CWR_STATUS_write(address);

     address = 3'b011;
     drive_data = 8'b11000010;
     // drive portC[4] to default 1
     drive_portC[4] = 1;
     // drive portC[6] to default 1
     drive_portC[6] = 1;
     CWR_STATUS_write(address);
     drive_data = 8'hzz;

     // read from portA
     address = 0;
     drive_portA = 8'ha5;
     drive_portC[4] = 0; // this is to have stbab at low
     #cycle;
```

```
      drive_portC[4] = 1; // this is to have stbab back at high
      read_port;
      #cycle;
      drive_portA = 8'hzz;

      // write to portA

      drive_data = 8'haa;
      write_port;
      #cycle;
      drive_portC[6] = 0; // this is to have ackab at low
      #cycle;
      drive_portC[6] = 1; // this is to have ackab back
      drive_portA = 8'hzz;
      #cycle;
      drive_data = 8'hzz;
      #cycle;

      // write to portB
      address = 1;
      drive_data = 8'h67;
      write_port;
      drive_data = 8'hzz;
      #cycle;


end


task write_port;
begin
      tb_wrb = 1;
      tb_rdb = 1;
      #cycle;
      tb_wrb = 0;
      #cycle;
      tb_wrb = 1;
      #cycle;
end
endtask

task read_port;
begin
      tb_wrb = 1;
      tb_rdb = 1;
      #cycle;
```

```
        tb_rdb = 0;
        #cycle;
        tb_rdb = 1;
        #cycle;
end
endtask

task CWR_STATUS_write;
input [2:0] address;
begin
        tb_reset = 0;
        tb_rdb = 1;
        tb_wrb = 1;
        #cycle;
        tb_wrb = 0;
        #cycle;
        tb_wrb = 1;
        #cycle;
end
endtask

task task_reset;
begin
        tb_reset = 0;
        #cycle;
        tb_reset = 1;
        #cycle;
        tb_reset = 0;
        #cycle;
end
endtask

ppi ppi_inst (.portA(tb_portA), .portB(tb_portB),
.portC(tb_portC), .rdb(tb_rdb), .wrb(tb_wrb),
.a2(tb_a2), .a1(tb_a1), .a0(tb_a0), .reset(tb_reset),
.data(tb_data));

endmodule
```

Referring to Figure 6.45:

a. In the circle marked 1, ***reset*** occurs to reset the ***CWR*** register to "9e" and the ***STATUS*** register to "0".
b. In the circle marked 2, ***wrb*** pulses low, causing a write to address "111" (***STATUS*** register). The data on ***data[7:0]*** bus is "0". This data is written

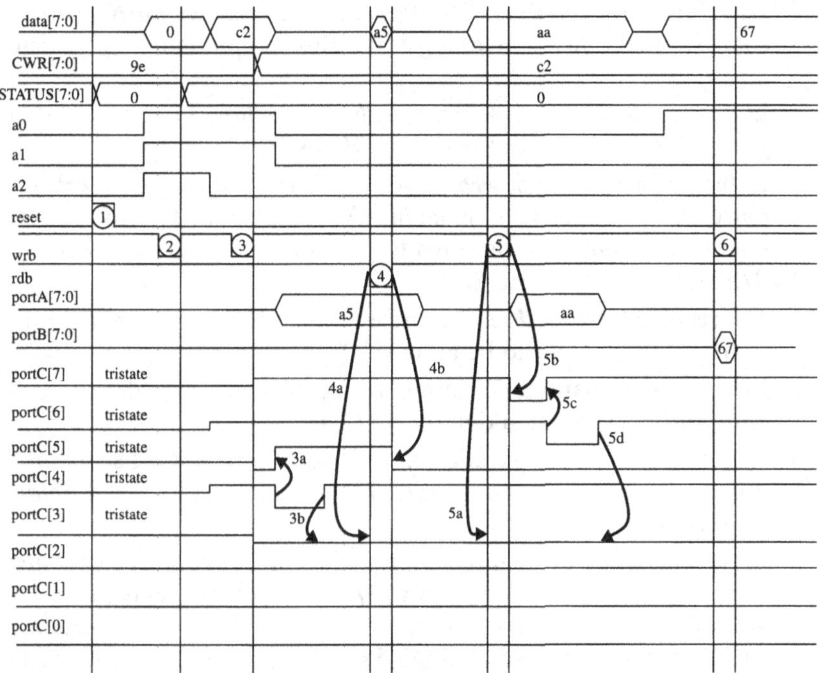

FIGURE 6.45. Diagram showing simulation results of test bench Example 6.13.

into the ***STATUS*** register. The contents of the ***STATUS*** register are used as a qualifier in generating the control signals ***portC*** when the PLB operates in mode 1 or mode 2 operation.

c. In the circle marked 3, ***wrb*** pulses low, causing a write to address "011" (***CWR*** register). The data on ***data[7:0]*** bus is "c2". This data is written into the ***CWR*** register. ("c2" in the ***CWR*** register configures the PLB to function in mode 2 operation with ***portA*** as a strobed I/O port and ***portB*** as an input port. ***PortC*** acts as the control signals for ***portA***.)

d. For the curved arrow marked 3a, ***portC[4]*** (input control signal ***stbab***) drives logic "0". This causes ***portC[5]*** (output control signal ***ibfa***) to be at logic "1".

e. For the curved arrow marked 3b, ***portC[4]*** (input control signal ***stbab***) drives logic "1". This causes ***portC[3]*** (output control signal ***intra***) to be at logic "1". However, because the qualifier for ***portC[3]*** is ***STATUS[5:4]*** (the content of ***STATUS[5:4]*** is "0"), ***portC[3]*** remains at logic "0" instead of toggling to logic "1".

f. In the circle marked 4, ***rdb*** pulses low. The falling edge of ***rdb*** causes ***portC[3]*** (output control signal ***intra***) to be at logic "0". This is represented by the curved arrow 4a. However, from Figure 6.45, ***portC[3]*** is already at logic "0" because the qualifier ***STATUS[5:4]*** is at logic "0".

g. The rising edge of ***rdb*** in the circle marked 4 causes ***portC[5]*** (output control signal ***ibfa***) to be at logic "0". This is represented by the curved arrow 4b.
h. In the circle marked 4, when ***rdb*** pulses low, the data ("a5") at ***portA[7:0]*** is read onto the ***data[7:0]*** bus ("a5").
i. In the circle marked 5, ***wrb*** pulses low. The falling edge of ***wrb*** causes ***portC[3]*** (output control signal ***intra***) to be at logic "0". However, from Figure 6.45, ***portC[3]*** is already at logic "0" because the qualifier ***STATUS[5:4]*** is at logic "0".
j. The rising edge of ***wrb*** in the circle marked 5 causes ***portC[7]*** (output control signal ***obfab***) to be at logic "0".
k. In the circle marked 5, during rising edge of ***wrb***, data ("aa") at ***data[7:0]*** bus is written onto ***portA[7:0]***.
l. In the curved arrow marked 5c, ***portC[6]*** (input control signal ***ackab***) drives logic "0". This causes ***portC[7]*** (output control signal ***obfab***) to be at logic "1".
m. In the curved arrow marked 5d, ***portC[6]*** (input control signal ***ackab***) drives logic "1". This causes ***portC[3]*** (output control signal ***intra***) to be at logic "1". However, because the qualifier for ***portC[3]*** is ***STATUS[5:4]*** (the content of ***STATUS[5:4]*** is "0"), ***portC[3]*** remains at logic "0" instead of toggling to logic "1".
n. In the circle marked 6, ***wrb*** pulses low, causing a write to address "001". Because the PLB is operating in mode 2 with ***portB*** as output, the data at ***data[7:0]*** bus ("67") is written to ***portB***.

Appendix

APPENDIX A.1 TWO-BIT BY TWO-BIT ADDER

Verilog code for a two-bit by two-bit adder:

```
module addition_2bit (inputA, inputB, outputA);

input [1:0] inputA, inputB;
output [2:0] outputA;

wire [2:0] outputA;

assign outputA = inputA + inputB;

endmodule
```

Verilog test bench to simulate the two-bit by two-bit adder:

```
module addition_2bit_tb ();

reg [1:0] reg_inputA, reg_inputB;
wire [2:0] wire_outputA;

integer i,j;

initial
begin
     for (i=0; i<4; i=i+1)
          begin
               reg_inputA = i;
```

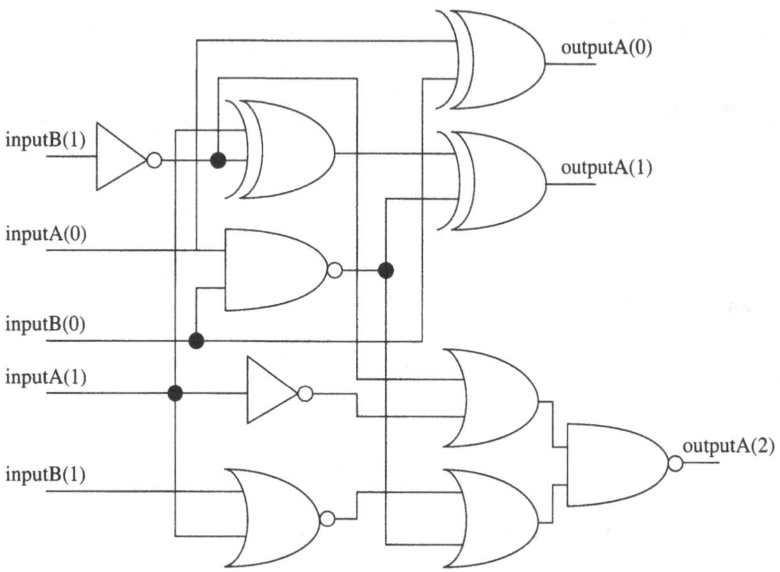

FIGURE A.1. Synthesized logic for the two-bit by two-bit adder.

```
                for (j=0; j<4; j=j+1)
                     begin
                          reg_inputB = j;
                          #10;
                     end
          end
end

addition_2bit addition_2bit_inst (.inputA(reg_inputA),
.inputB(reg_inputB), .outputA(wire_outputA));

initial
begin
     $monitor ("inputA %b%b inputB %b%b outputA
%b%b%b", reg_inputA[1], reg_inputA[0], reg_inputB[1],
reg_inputB[0],
     wire_outputA[2], wire_outputA[1],
wire_outputA[0]);
end

endmodule
```

Simulation results for simulating the two-bit by two-bit adder:

```
inputA 00 inputB 00 outputA 000
inputA 00 inputB 01 outputA 001
inputA 00 inputB 10 outputA 010
inputA 00 inputB 11 outputA 011
inputA 01 inputB 00 outputA 001
inputA 01 inputB 01 outputA 010
inputA 01 inputB 10 outputA 011
inputA 01 inputB 11 outputA 100
inputA 10 inputB 00 outputA 010
inputA 10 inputB 01 outputA 011
inputA 10 inputB 10 outputA 100
inputA 10 inputB 11 outputA 101
inputA 11 inputB 00 outputA 011
inputA 11 inputB 01 outputA 100
inputA 11 inputB 10 outputA 101
inputA 11 inputB 11 outputA 110
```

APPENDIX A.2 TWO-BIT BY TWO-BIT SUBTRACTOR

Verilog code for a two-bit by two-bit subtractor:

```
module subtraction_2bit (inputA, inputB, outputA);

input [1:0] inputA, inputB;
output [2:0] outputA;

wire [2:0] outputA;

assign outputA = inputA - inputB;

endmodule
```

Verilog test bench to simulate the two-bit by two-bit subtractor:

```
module subtraction_2bit_tb ();

reg [1:0] reg_inputA, reg_inputB;
wire [2:0] wire_outputA;

integer i,j;

initial
begin
     for (i=0; i<4; i=i+1)
```

```
        begin
            reg_inputA = i;
            for (j=0; j<4; j=j+1)
                begin
                    reg_inputB = j;
                    #10;
                end
        end
end

subtraction_2bit subtraction_2bit_inst
(.inputA(reg_inputA), .inputB(reg_inputB),
.outputA(wire_outputA));

initial
begin
    $monitor ("inputA %h inputB %h outputA %h",
reg_inputA[1:0], reg_inputB[1:0], wire_outputA[2:0]);
end

endmodule
```

Simulation results for simulating the two-bit by two-bit subtractor:

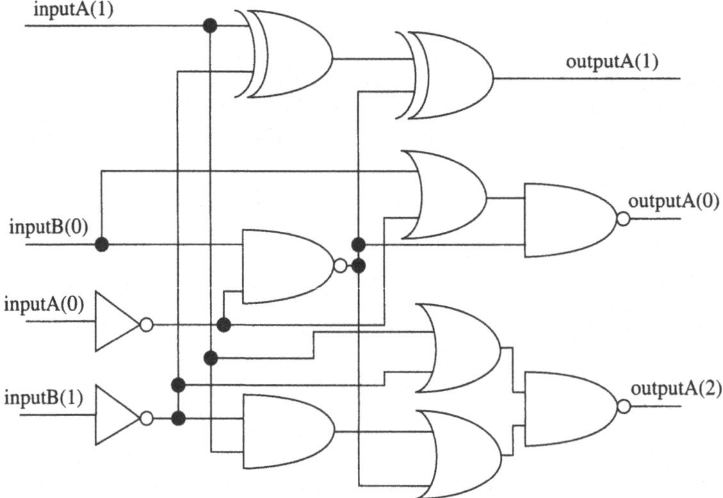

FIGURE A.2. Synthesized logic for the two-bit by two-bit subtractor.

```
inputA 0 inputB 0 outputA 0
inputA 0 inputB 1 outputA 7
inputA 0 inputB 2 outputA 6
inputA 0 inputB 3 outputA 5
inputA 1 inputB 0 outputA 1
inputA 1 inputB 1 outputA 0
inputA 1 inputB 2 outputA 7
inputA 1 inputB 3 outputA 6
inputA 2 inputB 0 outputA 2
inputA 2 inputB 1 outputA 1
inputA 2 inputB 2 outputA 0
inputA 2 inputB 3 outputA 7
inputA 3 inputB 0 outputA 3
inputA 3 inputB 1 outputA 2
inputA 3 inputB 2 outputA 1
inputA 3 inputB 3 outputA 0
```

APPENDIX A.3 FOUR-BIT BY FOUR-BIT MULTIPLIER

Verilog code for a four-bit by four-bit multiplier:

```
module multiplication_4bit (inputA, inputB, outputA);

input [3:0] inputA, inputB;
output [7:0] outputA;

wire [7:0] outputA;

assign outputA = inputA * inputB;

endmodule
```

Verilog test bench to simulate the four-bit by four-bit multiplier:

```
module multiplication_tb ();

reg [3:0] reg_inputA, reg_inputB;

wire [7:0] wire_outputA;

integer i,j;

initial
begin
```

```
    for (i=0; i<16; i=i+1)
         begin
              reg_inputA = i;
              for (j=0; j<16; j=j+1)
                   begin
                        reg_inputB = j;
                        #10;
                   end
         end
end

multiplication_4bit multiplication_4bit_inst
(.inputA(reg_inputA), .inputB(reg_inputB),
.outputA(wire_outputA));

initial
begin
    $monitor ("inputA %h inputB %h outputA %h",
reg_inputA, reg_inputB, wire_outputA);
end

endmodule
```

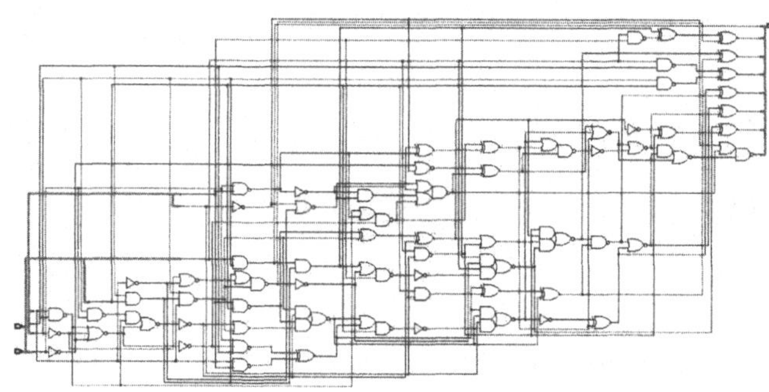

FIGURE A.3. Synthesized logic for the four-bit by four-bit multiplier.

Simulation results for simulating the four-bit by four-bit multiplier:

```
inputA 0 inputB 0 outputA 00
inputA 0 inputB 1 outputA 00
inputA 0 inputB 2 outputA 00
inputA 0 inputB 3 outputA 00
inputA 0 inputB 4 outputA 00
inputA 0 inputB 5 outputA 00
inputA 0 inputB 6 outputA 00
inputA 0 inputB 7 outputA 00
inputA 0 inputB 8 outputA 00
inputA 0 inputB 9 outputA 00
inputA 0 inputB a outputA 00
inputA 0 inputB b outputA 00
inputA 0 inputB c outputA 00
inputA 0 inputB d outputA 00
inputA 0 inputB e outputA 00
inputA 0 inputB f outputA 00
inputA 1 inputB 0 outputA 00
inputA 1 inputB 1 outputA 01
inputA 1 inputB 2 outputA 02
inputA 1 inputB 3 outputA 03
inputA 1 inputB 4 outputA 04
inputA 1 inputB 5 outputA 05
inputA 1 inputB 6 outputA 06
inputA 1 inputB 7 outputA 07
inputA 1 inputB 8 outputA 08
inputA 1 inputB 9 outputA 09
inputA 1 inputB a outputA 0a
inputA 1 inputB b outputA 0b
inputA 1 inputB c outputA 0c
inputA 1 inputB d outputA 0d
inputA 1 inputB e outputA 0e
inputA 1 inputB f outputA 0f
inputA 2 inputB 0 outputA 00
inputA 2 inputB 1 outputA 02
inputA 2 inputB 2 outputA 04
inputA 2 inputB 3 outputA 06
inputA 2 inputB 4 outputA 08
inputA 2 inputB 5 outputA 0a
inputA 2 inputB 6 outputA 0c
inputA 2 inputB 7 outputA 0e
inputA 2 inputB 8 outputA 10
inputA 2 inputB 9 outputA 12
inputA 2 inputB a outputA 14
```

```
inputA 2 inputB b outputA 16
inputA 2 inputB c outputA 18
inputA 2 inputB d outputA 1a
inputA 2 inputB e outputA 1c
inputA 2 inputB f outputA 1e
inputA 3 inputB 0 outputA 00
inputA 3 inputB 1 outputA 03
inputA 3 inputB 2 outputA 06
inputA 3 inputB 3 outputA 09
inputA 3 inputB 4 outputA 0c
inputA 3 inputB 5 outputA 0f
inputA 3 inputB 6 outputA 12
inputA 3 inputB 7 outputA 15
inputA 3 inputB 8 outputA 18
inputA 3 inputB 9 outputA 1b
inputA 3 inputB a outputA 1e
inputA 3 inputB b outputA 21
inputA 3 inputB c outputA 24
inputA 3 inputB d outputA 27
inputA 3 inputB e outputA 2a
inputA 3 inputB f outputA 2d
inputA 4 inputB 0 outputA 00
inputA 4 inputB 1 outputA 04
inputA 4 inputB 2 outputA 08
inputA 4 inputB 3 outputA 0c
inputA 4 inputB 4 outputA 10
inputA 4 inputB 5 outputA 14
inputA 4 inputB 6 outputA 18
inputA 4 inputB 7 outputA 1c
inputA 4 inputB 8 outputA 20
inputA 4 inputB 9 outputA 24
inputA 4 inputB a outputA 28
inputA 4 inputB b outputA 2c
inputA 4 inputB c outputA 30
inputA 4 inputB d outputA 34
inputA 4 inputB e outputA 38
inputA 4 inputB f outputA 3c
inputA 5 inputB 0 outputA 00
inputA 5 inputB 1 outputA 05
inputA 5 inputB 2 outputA 0a
inputA 5 inputB 3 outputA 0f
inputA 5 inputB 4 outputA 14
inputA 5 inputB 5 outputA 19
inputA 5 inputB 6 outputA 1e
inputA 5 inputB 7 outputA 23
```

```
inputA 5 inputB 8 outputA 28
inputA 5 inputB 9 outputA 2d
inputA 5 inputB a outputA 32
inputA 5 inputB b outputA 37
inputA 5 inputB c outputA 3c
inputA 5 inputB d outputA 41
inputA 5 inputB e outputA 46
inputA 5 inputB f outputA 4b
inputA 6 inputB 0 outputA 00
inputA 6 inputB 1 outputA 06
inputA 6 inputB 2 outputA 0c
inputA 6 inputB 3 outputA 12
inputA 6 inputB 4 outputA 18
inputA 6 inputB 5 outputA 1e
inputA 6 inputB 6 outputA 24
inputA 6 inputB 7 outputA 2a
inputA 6 inputB 8 outputA 30
inputA 6 inputB 9 outputA 36
inputA 6 inputB a outputA 3c
inputA 6 inputB b outputA 42
inputA 6 inputB c outputA 48
inputA 6 inputB d outputA 4e
inputA 6 inputB e outputA 54
inputA 6 inputB f outputA 5a
inputA 7 inputB 0 outputA 00
inputA 7 inputB 1 outputA 07
inputA 7 inputB 2 outputA 0e
inputA 7 inputB 3 outputA 15
inputA 7 inputB 4 outputA 1c
inputA 7 inputB 5 outputA 23
inputA 7 inputB 6 outputA 2a
inputA 7 inputB 7 outputA 31
inputA 7 inputB 8 outputA 38
inputA 7 inputB 9 outputA 3f
inputA 7 inputB a outputA 46
inputA 7 inputB b outputA 4d
inputA 7 inputB c outputA 54
inputA 7 inputB d outputA 5b
inputA 7 inputB e outputA 62
inputA 7 inputB f outputA 69
inputA 8 inputB 0 outputA 00
inputA 8 inputB 1 outputA 08
inputA 8 inputB 2 outputA 10
inputA 8 inputB 3 outputA 18
inputA 8 inputB 4 outputA 20
```

```
inputA 8 inputB 5 outputA 28
inputA 8 inputB 6 outputA 30
inputA 8 inputB 7 outputA 38
inputA 8 inputB 8 outputA 40
inputA 8 inputB 9 outputA 48
inputA 8 inputB a outputA 50
inputA 8 inputB b outputA 58
inputA 8 inputB c outputA 60
inputA 8 inputB d outputA 68
inputA 8 inputB e outputA 70
inputA 8 inputB f outputA 78
inputA 9 inputB 0 outputA 00
inputA 9 inputB 1 outputA 09
inputA 9 inputB 2 outputA 12
inputA 9 inputB 3 outputA 1b
inputA 9 inputB 4 outputA 24
inputA 9 inputB 5 outputA 2d
inputA 9 inputB 6 outputA 36
inputA 9 inputB 7 outputA 3f
inputA 9 inputB 8 outputA 48
inputA 9 inputB 9 outputA 51
inputA 9 inputB a outputA 5a
inputA 9 inputB b outputA 63
inputA 9 inputB c outputA 6c
inputA 9 inputB d outputA 75
inputA 9 inputB e outputA 7e
inputA 9 inputB f outputA 87
inputA a inputB 0 outputA 00
inputA a inputB 1 outputA 0a
inputA a inputB 2 outputA 14
inputA a inputB 3 outputA 1e
inputA a inputB 4 outputA 28
inputA a inputB 5 outputA 32
inputA a inputB 6 outputA 3c
inputA a inputB 7 outputA 46
inputA a inputB 8 outputA 50
inputA a inputB 9 outputA 5a
inputA a inputB a outputA 64
inputA a inputB b outputA 6e
inputA a inputB c outputA 78
inputA a inputB d outputA 82
inputA a inputB e outputA 8c
inputA a inputB f outputA 96
inputA b inputB 0 outputA 00
inputA b inputB 1 outputA 0b
```

```
inputA b inputB 2 outputA 16
inputA b inputB 3 outputA 21
inputA b inputB 4 outputA 2c
inputA b inputB 5 outputA 37
inputA b inputB 6 outputA 42
inputA b inputB 7 outputA 4d
inputA b inputB 8 outputA 58
inputA b inputB 9 outputA 63
inputA b inputB a outputA 6e
inputA b inputB b outputA 79
inputA b inputB c outputA 84
inputA b inputB d outputA 8f
inputA b inputB e outputA 9a
inputA b inputB f outputA a5
inputA c inputB 0 outputA 00
inputA c inputB 1 outputA 0c
inputA c inputB 2 outputA 18
inputA c inputB 3 outputA 24
inputA c inputB 4 outputA 30
inputA c inputB 5 outputA 3c
inputA c inputB 6 outputA 48
inputA c inputB 7 outputA 54
inputA c inputB 8 outputA 60
inputA c inputB 9 outputA 6c
inputA c inputB a outputA 78
inputA c inputB b outputA 84
inputA c inputB c outputA 90
inputA c inputB d outputA 9c
inputA c inputB e outputA a8
inputA c inputB f outputA b4
inputA d inputB 0 outputA 00
inputA d inputB 1 outputA 0d
inputA d inputB 2 outputA 1a
inputA d inputB 3 outputA 27
inputA d inputB 4 outputA 34
inputA d inputB 5 outputA 41
inputA d inputB 6 outputA 4e
inputA d inputB 7 outputA 5b
inputA d inputB 8 outputA 68
inputA d inputB 9 outputA 75
inputA d inputB a outputA 82
inputA d inputB b outputA 8f
inputA d inputB c outputA 9c
inputA d inputB d outputA a9
inputA d inputB e outputA b6
```

```
inputA d inputB f outputA c3
inputA e inputB 0 outputA 00
inputA e inputB 1 outputA 0e
inputA e inputB 2 outputA 1c
inputA e inputB 3 outputA 2a
inputA e inputB 4 outputA 38
inputA e inputB 5 outputA 46
inputA e inputB 6 outputA 54
inputA e inputB 7 outputA 62
inputA e inputB 8 outputA 70
inputA e inputB 9 outputA 7e
inputA e inputB a outputA 8c
inputA e inputB b outputA 9a
inputA e inputB c outputA a8
inputA e inputB d outputA b6
inputA e inputB e outputA c4
inputA e inputB f outputA d2
inputA f inputB 0 outputA 00
inputA f inputB 1 outputA 0f
inputA f inputB 2 outputA 1e
inputA f inputB 3 outputA 2d
inputA f inputB 4 outputA 3c
inputA f inputB 5 outputA 4b
inputA f inputB 6 outputA 5a
inputA f inputB 7 outputA 69
inputA f inputB 8 outputA 78
inputA f inputB 9 outputA 87
inputA f inputB a outputA 96
inputA f inputB b outputA a5
inputA f inputB c outputA b4
inputA f inputB d outputA c3
inputA f inputB e outputA d2
inputA f inputB f outputA e1
```

Glossary

APR auto place and route.

ASIC application-specific integrated circuit.

Back annotation back annotating physical layout information from layout to design to enable a more accurate simulation.

Contention more than one signal is driving a node.

Design compiler a synthesis tool from Synopsys.

EDA electronic design automation.

FPGA field programmable gate array.

HDL hardware description language.

Hold time time required for a signal to be held valid with reference to clock change.

Mentor Graphics an EDA tool company.

Parasitics resistance and capacitance that is caused by layout routing.

RTL register transfer level.

Schematic capture a method of design where circuits are hand drawn.

Sensitivity list a list that contains all the signals that will invoke the corresponding process.

Setup time time required for a signal to be held valid prior to a clock change.

Simulation using a set of input stimulus to verify the functionality of a design.

Summit Design an EDA tool company.

Synopsys an EDA tool company.

Synthesis the process of conversion from HDL to logic gates using a synthesis tool.

Synthesizable code HDL code that is coded in a manner that allows it to be synthesized.

Test bench a wraparound on a design that injects stimulus into the "design under test" to verify the functionality of the design.

UDP user-defined primitive.

Verilog an HDL language.

VHDL another HDL language.

VLSI very-large-scale integration.

Bibliography

VHDL Coding and Logic Synthesis with Synopsys, by Weng Fook Lee, Academic Press, 2000.

Verilog HDL: A Guide to Digital Design and Synthesis, by Samir Palnitkar, Prentice Hall, 1994.

A Verilog HDL Primer 2nd Ed., by J. Bhasker, Star Galaxy Publishing, 1999.

Index

9 780471 429760